Complementarity, Causality, and Explanation

Complementarity, Causality, and Explanation

John Losee

Routledge
Taylor & Francis Group
LONDON AND NEW YORK

First published 2013 by Transaction Publishers

2 Park Square, Milton Park, Abingdon, Oxfordshire OX14 4RN
711 Third Avenue, New York, NY 10017

Routledge is an imprint of the Taylor & Francis Group, an informa business

First issued in paperback 2017

Library of Congress Catalog Number: 2012027390

Library of Congress Cataloging-in-Publication Data

Losee, John.
Complementarity, causality, and explanation / John Losee.
p. cm.
ISBN 978-1-4128-4990-6
1. Causality (Physics) 2. Complementarity (Physics) 3. Quantum theory. 4. Quantum theory--Religious aspects--Christianity. 5. Bohr, Niels, 1885-1962. I. Title.
QC6.4.C3L67 2013
530.01--dc23

2012027390

ISBN 13: 978-1-4128-4990-6 (hbk)
ISBN 13: 978-1-138-50836-1 (pbk)

Contents

Introduction

Philosophers of science from Aristotle to Mill have disagreed about the nature of the relationship between cause and effect. Three influential positions were

1. cause → effect is a logical relationship, for instance,
    a. cause c is a necessary condition of effect e ($e \therefore c$), or
    b. cause c is a sufficient condition of effect e ($c \therefore e$);
2. cause → effect is a relation between events; and
3. cause → effect is a relation between successive states of physical systems.

Despite disagreement about the nature of causal relatedness, philosophers of science from Aristotle to Mill were in agreement that successful causal attribution explains why an effect takes place (or is what it is). If one can establish that e is the effect of cause c, then one has explained why e occurred (or is what it is). I shall refer to this position as the "causality implies explanation" thesis.

The "causality implies explanation" thesis was unchallenged throughout the nineteenth century. It received support from developments in physics and chemistry during this period. Scientists increasingly came to view causal relatedness as the regularity of sequence between successive states of (relatively isolated) physical systems. If a regularity is established between successive states of a physical system (e.g., a charged body moving in an electromagnetic field) then the realization of state B is explained by pointing out the prior realization of state A, namely,

$$\begin{array}{c} (x)\left[Ax \supset Bx\right] \\ Aa \\ \hline \therefore Ba \end{array}$$

where Aa = state A is realized in instance a.

The "causality implies explanation" thesis received a substantial challenge from the development of quantum mechanics. The noted historian of science I. Bernard Cohen declared that "quantum theory has . . . produced a fundamental change in our scientific thinking and philosophy of science."[1]

Regularity between states holds for quantum-systems-as-observed, but satisfactory explanations of quantum phenomena are unavailable. Quantum theory created a rift between causality and explanation.

Among the responses offered upon recognition of this rift are

1. Restrict the "causality implies explanation" thesis to the results of measurements made of quantum systems;
2. Restore the "causality implies explanation" thesis by locating causal relatedness at a deeper level of interaction that underlies and explains quantum phenomena, much as kinetic molecular theory underlies and explains the macroscopic behavior of gases; and
3. Superimpose on the formalism of quantum theory mutually exclusive, but complementary, pictures that depict
   a. causal interactions and spatiotemporal descriptions, or
   b. motions of classical particles and classical waves.

Response 3 is the resolution recommended by Niels Bohr. Bohr's complementary pictures apply to quantum systems between observations made of them. Bohr maintained that they have explanatory value, despite the prohibition against combining a statement of causal interaction with spatiotemporal description (or a particle picture with a wave picture) for a quantum-system-observed-under-a-specific-experimental-arrangement.

Bohr sought to extend the principle of complementarity that he had developed to provide explanations of quantum phenomena to the disciplines of biology and psychology. In biology, the investigator must elect one of two mutually exclusive experimental orientations: perform dissections and chemical analyses on an organism, or observe the organism's adaptive responses in its environment. In psychology, the investigator must elect to pursue either causal analyses or vitalistic interpretations.

Bohr insisted that *complete* explanations in biology and psychology utilize mutually exclusive, but complementary, alternatives. Moreover, he took the presumed role of complementarity in domains other than quantum physics to provide enhanced support for the principle of complementarity itself. Bohr was committed to the unity of empirical knowledge, a unity to be achieved, in part, by applications of the principle of complementarity.

Bohr did not, but well might have, sought to apply the principle of complementarity to Christian religious discourse as well. This discourse includes both claims about the immutable attributes of God and claims about a dialog between God and man. These types of religious linguistic usage are mutually exclusive. It might be suggested, however, that these two types of linguistic usage are complementary. If this suggestion has merit, then theology could be added to the list of disciplines whose explanations are derived from applications of the principle of complementarity.

The adequacy of complementarity as a methodological principle for biology, psychology, and theology, as well as for quantum physics itself, remains to be assessed. An attempt to do so is a feature of the latter chapters of this work.

Plan of the Work

1. Survey important theories of causality from Aristotle to Mill.
2. Examine these theories to establish that there was general support for the thesis that "causality implies explanation."
3. Show how nineteenth-century theories in physics and chemistry were informed by, and in turn provided support for, the dominant theory that causality is regularity between successive states of physical systems.
4. Trace early nineteenth-century developments in physics that provided the background for the emergence of quantum theory.
5. Make clear how the Copenhagen interpretation of quantum phenomena requires a division into three levels of language—observational, theoretical, and pictorial.
6. Indicate problems created for the "causality implies explanation" thesis by the emergence of quantum theory.
7. Examine responses made in response to recognition of the rift between causality and explanation in quantum mechanics, among them Bohr's appeal to complementary descriptions at a pictorial level of interpretation.
8. Present and evaluate Bohr's proposal to apply the principle of complementarity to the disciplines of biology and psychology.
9. Test the feasibility of using complementarity as a methodological principle for interpretations within Christian theology.
10. Assess whether applications of the principle of complementarity are necessary for explanations of phenomena in the quantum domain itself.

Note

1. I. Bernard Cohen, *Revolution in Science* (Cambridge, MA: Harvard University Press, 1985), 420.

1

The Causality-Explanation Connection from Aristotle to Mill

From Aristotle to Bohr, there was general agreement that if a causal attribution is correct, then this explains why the effect in question took place. If "*c* caused *e*" is true, then "the occurrence of *c* explains why *e* occurred" is true as well. Successful causal attribution is a sufficient condition of scientific explanation.

I shall refer to this claim as the "causality implies explanation" thesis. It is a conditional claim that asserts that the identification of a cause provides an explanation for the occurrence of its effect.

The converse of the "causality implies explanation" thesis is not true. Causal attribution is not a necessary condition of scientific explanation. Applications of Archimedes's principle of buoyancy, Newton's axiom of inertial motion, or Boyle's law achieve explanatory success without establishing causal relatedness.

Aristotle

Aristotle (384–322 BC) was the first to defend the thesis that specifying a causal relation is a sufficient condition of scientific explanation. He maintained that one achieves understanding of a process by uncovering its causes. Aristotle declared that

> we suppose ourselves to possess unqualified scientific knowledge . . . when we think that we know the cause on which the fact depends, as the cause of that fact and no other, and, further, that the fact could not be other than it is.[1]

Aristotle was overly optimistic about what science can achieve. One cannot prove that a fact cannot be other than it is. Isaac Newton was correct to insist that the interpretations of science cannot achieve the status of necessary truths.[2] Nevertheless, Aristotle's position that successful causal inquiry counts as scientific explanation may be accepted without commitment to the unattainable ideal of necessary empirical truth.

Aristotle held that if a causal relation is specified, then a corresponding scientific explanation is available, but the converse is not true. He noted that one may formulate noncausal explanations of the existence and natures of substances.[3]

Aristotle viewed natural processes from two perspectives: as an imposition of form upon matter (e.g., ice is created from water) or as a transition from potentiality to actuality (e.g., a tadpole evolves into a frog). A scientist establishes the cause of a process by discovering the nature of an impressed form or the nature of a potentiality that has become actualized.

Aristotle discussed four types of causal factors that are involved in a natural process: (1) the form of its development (the *formal* cause), (2) its material embodiment (the *material* cause), (3) the agency responsible (the *efficient* cause), and (4) the end achieved (the *final* cause).[4] Suppose a cuttlefish emits ink as a shark swims past. The formal cause is the pattern that correlates ink emission and nearby motion; the material cause is the particular composition of the ink; the efficient cause is the mechanism in the cuttlefish that expels ink in response to perceived input from its environment; and the final cause is to achieve camouflage and, ultimately, self-preservation.

Aristotle sometimes held that a complete scientific explanation requires specification of all four types of cause. On other occasions, he conceded that two or more types of cause often coincide.[5] In his own scientific investigations, Aristotle often restricted inquiry to just one or two of the four types of cause.

Aristotle maintained that a cause produces its effect, and to cite a cause is to explain why its effect occurs. To establish the cause of a process (qua transition from potentiality to actuality) is to explain why the process is what it is.

Aristotle required that a successful scientific explanation take the form of a deductive argument whose premises state the cause of the attribution asserted in the conclusion. The argument takes the form of a syllogism, namely,

$$\frac{\begin{array}{c}\text{All } M \text{ are } P;\\ \text{All } S \text{ are } M;\end{array}}{\therefore \text{All } S \text{ are } M}$$

where *S*, *P*, and *M* represent classes related as follows:

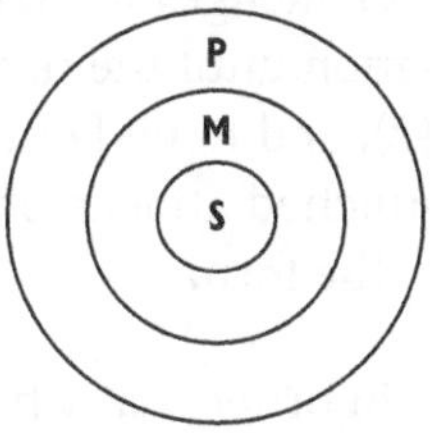

Aristotle cautioned that instantiation of the appropriate deductive argument form with true premises does not guarantee that scientific explanation has been achieved. Consider the argument:

All objects that do not twinkle are objects near the earth.
All planets are objects that do not twinkle.
∴ All planets are objects near the earth.[6]

In this syllogism, middle term *M* (objects that do not twinkle) does not establish causal relatedness between planets and objects near the earth. By contrast, the following syllogism does qualify as a scientific explanation:

All objects near the earth are objects that do not twinkle.
All planets are objects near the earth.
∴ All planets are objects that do not twinkle.

Aristotle maintained that the premises of this argument do state the cause of the fact that planets do not twinkle.[7]

An important question is how a scientist is to determine whether the premises of a syllogistic argument state the cause of the attribution stated in the conclusion. Aristotle held that a conclusion states a causal relation only if the predicate term is true of the subject term in every instance of the subject, is true of the subject precisely and not as part of a larger whole, and is essential to the subject. Aristotle's analysis is not very helpful. He was unable to specify a general criterion of essential attribution. He did, however, provide illustrations, one of which is that slitting the throat of an animal is essentially related to its death.[8]

Bacon, Galileo, and Descartes

There were relatively few challenges to Aristotle's position on causation and scientific explanation until the time of Francis Bacon (1561–1626), Galileo (1564–1642), and Rene Descartes (1596–1650). Bacon, Galileo, and Descartes launched attacks on Aristotle's appeal to final causes. A final cause takes the form.

x occured in order that y be achieved.

Aristotle had claimed a falling body approaches the surface of the earth in order to reach its "natural place," and fire rises in order to reach its "natural place" (a spherical shell inside the orbit of the moon).

To take seriously the operation of final causes in nature is to grant to the future a causal influence on the present. Bacon, Galileo, and Descartes sought to exclude reference to final causes from scientific explanations.

Bacon maintained that the introduction of final causes for physical and biological phenomena serves only to promote verbal disputes that retard scientific progress.[9] He recommended that final causes be restricted to human volitional activity.

Galileo agreed. He restricted the causes of physical phenomena to factors present when the effects are present. A cause is, he wrote, "that which is always present when the effect is seen, and in whose absence the effect does not take place."[10]

In Galileo's view, c is the cause of e only if its presence is necessary for the occurrence of e. Whether or not some end is achieved by a transition is irrelevant. Galileo also banned Aristotle's final causes from the arena of scientific inquiry.

Descartes was more specific. He sought to restrict scientific inquiry to what Aristotle had called "efficient causes."[11] Final causes are relics of a prescientific age. However, Descartes had no quarrel with Aristotle's claim that to identify a cause is to explain why its effect is what it is. Indeed, Descartes made the further claim that *only* causal accounts qualify as scientific explanations.[12]

Descartes was convinced that God created the world in a single act of creation. God, of course, is a Perfect Being. Descartes believed that he could prove that a Perfect Being would not create a universe in which there exist spaces devoid of matter. Since no vacuum is possible, all motion in the universe must be the result of contact action.

Hence, a cause must be an occurrence of impact or pressure. No action-at-a-distance can have causal status.[13] Consequently, all causation is efficient causation, and if an efficient cause is established, then a causal explanation is available.

But then how to explain the motions of the planets around the sun? There is no visible matter pressing inward on the planets. Nevertheless, the planets do describe discrete orbits around the sun. Descartes suggested that an invisible vortex of ethereal particles carries the planets around the sun, much as bits of cork are carried around the center of a whirlpool.[14] He maintained that because the pressure exerted by the vortex is a bona fide cause, a scientific explanation of planetary motions has been given.

Descartes's followers complained that Newton's attribution of planetary orbits to gravitational attraction from the sun fails to explain why these orbits exist. To show that planetary orbits conform to the results of calculations that apply Newton's three axioms of motion and the law of gravitational attraction is merely to "save the appearances." A *genuine* explanation requires specification of the cause of orbital motions, and this, in turn, requires specification of bodies that exert inward-directed pressure on the planets.

Newton, in response, presented a devastating criticism of Descartes's vortex theory.[15] He observed that if a planet were carried around the sun by an invisible ethereal vortex, then the density of the vortex at a planet's distance from the sun must be equal to that of the planet itself. If the density of the vortex at the planet's distance from the sun were greater than the density of the planet, then the planet would spiral inward toward the sun. If the density of the vortex at the planet's distance from the sun were less than that of the planet, then the planet would recede from its orbit. Given the observed density of the earth, the above requirement on the density of the vortex to assure orbital stability is implausible.

Nevertheless, Newton was willing to grant to Descartes a vortex of the required density because he could prove that the period of revolution of a planet within such a vortex would be proportional to the square of its distance from the sun. Since Johannes Kepler had shown that the period of a planet is proportional to the three-halves power of its distance from the sun, Descartes's vortex theory is false.

This would seem to be game, set, and match to Newton. However, Descartes's followers continued to insist that only the pressure of material particles could constrain a planet to persevere in its orbit around the sun. Leibniz and de Molieres introduced modified vortex theories in which multiple vortices combine to keep planets in their orbits.[16] These theories required vortices of ethereal particles to pass through each other without any interaction between them.

Perhaps the most extreme example of commitment to the thesis that "all causation is action by contact" is Christiaan Huygens' explanation of freely falling bodies. According to Huygens, an unsupported body falls to the earth because the vortex material at its upper surface presses upon it with greater force than does the vortex material on its lower surface. This force imbalance is the result of the greater linear velocity of the vortex material at the top surface of the body. To account for the observed acceleration of bodies in free fall (32 ft./sec.^{2}), Huygens required the vortex to revolve around the center of the earth with an angular velocity seventeen times greater than the angular velocity of the rotating earth.

It gets worse, however. Bodies released above the earth fall to the earth at all points on the earth's surface, at the poles as well as at the equator. To account for this, Huygens postulated the existence of a separate revolving vortex, centered on the center of the earth, for each location on the earth's surface. In this theory, for example, there is a revolving vortex above the Empire State Building and a revolving vortex above the Washington Monument. These vortices are required to intersect without affecting each other.

Huygens's model of vortices to account for free fall is one ugly theory. It requires that indefinitely many invisible whirlpools of ether revolve at extremely high velocities around the earth's center without interacting with one another. Its principal virtue is that it avoids reference to "action at a distance." From the Cartesian perspective, theories that invoke "action at a distance" are noncausal, and hence nonexplanatory.

Newton

Newton conceded to the Cartesians that it is one thing to develop mathematical relations—like the three axioms of motion and the law of universal gravitational attraction—from which planetary orbits may be calculated, and quite another thing to specify the cause of planetary motions. Newton usually maintained an agnostic position on the cause of the motions that conform to his laws of motion. He did not claim to have discovered the cause of gravitational attraction, for example.

However, he was willing to entertain—as a query to direct subsequent research—the hypothesis that an ethereal medium does cause the planets to remain in fixed orbits. Of course, this ethereal medium cannot be the Cartesian vortex which is inconsistent with Kepler's laws. Newton mused

> is not this Medium much rarer within the dense Bodies of the Sun, Stars, Planets and Comets, than in the empty celestial Spaces between them? And in passing from them to great distances, doth it not grow denser and denser perpetually, and thereby cause the gravity of those great Bodies towards one another, and of their parts towards the Bodies; every Body endeavouring to go from the denser parts of the Medium towards the rarer?[17]

Such a medium would have to be both exceedingly rare and extremely elastic. These are antithetical requirements.

Newton protected himself from criticism by maintaining that he merely was introducing a potential research program for others to pursue. The fact that he did present this query at all is testimony to the seductive power of the Cartesian position that a cause can operate only by impact or pressure.

Hume

David Hume (1711–76) was not interested in speculations about ethereal particles that fail to impress themselves on the senses. He subscribed to the empirical principle that all knowledge about the world is derived from, and is contained within, sense impressions. People experience regularities within their sense impressions. One sees a man release a ball from his hand, and then one sees the ball half-submerged in the mud below. But there is no sense impression of a glue that binds together these two events. It is a mistake to view causes as producers of their effects. Consistent with the empiricist principle, Hume suggested that regularity of sequence is a sufficient condition of causal relatedness. This position is sometimes spoken of as Hume's official position on causation.

On Hume's official position, a causal relation, objectively considered, is nothing over and above the constant sequential conjunction of the members of two classes of events. In addition, he maintained that, subjectively considered, a causal relation is a constant sequential conjunction such that, upon appearance of an event of the first class, the mind anticipates the occurrence of an event of the second class. Hume declared that

> when we look about us at towards external objects, and consider the operation of causes, we are never able, in a single instance, to discover any power or necessary connection; any quality which binds the effect to the cause, and renders the one an infallible consequence of the other. We only find, that the one does actually, in fact, follow the other.[18]

Unfortunately, Hume also made other claims about causal relatedness. He sometimes maintained that a causal relation is a constant sequential conjunction that does exhibit a necessary connection between the members of the two classes of events.[19] Moreover, Hume also proposed a counterfactual conditional interpretation of causal relatedness; actually occurring event *c* is a cause of event *e* if, and only if, if *c* had not taken place, then *e* would not have occurred either.[20]

Most commentators agree that the regularity view is the position most favored by Hume. Given this official position, does the discovery of a regularity between the members of two classes of events provide an explanation of why an effect event follows a cause event? Clearly it does not if explanation requires proof that the association is necessary. However, according to Hume, establishing a causal relation does achieve explanation in the sense that it provides all the knowledge that can be achieved from experience about the association of events in question.

Hume's official position fails as a definition of causal relatedness. It qualifies as causal correlations that are merely accidental. Suppose one placed two pendulum clocks side by side. Each tick of clock #1 may be followed by a tick of clock #2, but this is a merely accidental correlation. Other examples are easy to find. For instance, the four o'clock whistle at Factory *A* has been followed invariably by an egress of workers from Factory *B* across town; however, these two types of events are not causally related.

Hume could exclude such accidental correlations by appealing to his own counterfactual conditional definition of cause. Had the whistle at Factory *A* malfunctioned, there still would have been an egress of workers from Factory *B*. The whistle-egress correlation fails the counterfactual test for causal relatedness.

As a definition of cause, Hume's regularity view is too inclusive. Nevertheless, the regularity view gained many adherents during the nineteenth century. Many physical scientists, in particular, believed that the regularity view was superior to the view that causes somehow produce their effects.

Kant

Immanuel Kant (1724–1804) gave credit to Hume for awakening him from a "dogmatic slumber."[21] Kant was keenly aware that in Hume's official position, Newton's axioms of motion and Euclid's axioms of geometry are just de facto correlations. But surely Newton's axioms and the law of universal gravitational attraction have a necessity lacking in merely empirical generalizations such as "copper expands on heating" and "moss grows more luxuriantly on the north side of trees in the Northern Hemisphere." Kant insisted that Hume's official position on causality is too restrictive. Newtonian mechanics and Euclidean geometry express necessary truths about the universe.

One may question how this is possible if all knowledge about the world arises out of, and is given in, sense impressions. Kant's answer is that this thesis of empiricism is false. Although it is true that all knowledge about the world arises out of sense impressions, it is not true that all such knowledge is given in sense impressions.[22] He insisted that "we can *a priori* and prior to all given objects have a knowledge of those conditions on which alone experience of them is possible."[23]

Kant maintained that the knowing subject itself contributes empirically significant principles that create and organize knowledge about the world. One such principle is causality. According to Kant, it is application of the principle of causality that enables us to differently categorize

1. the successive perceptions of a log floating downstream, and
2. the successive perceptions of the façade of a building.

In the case of the log, separate perceptions follow one another according to a rule; in the case of the building façade the order of perceptions is arbitrary.

Suppose someone places a stone in the path of the sun's rays and finds that it has become warm. This is a sequence of perceptions. This presentation becomes cognitively significant insofar as it is an instantiation of the category of causality, via premise 2 below:[24]

1. First the stone is placed in the sun and then it becomes warm, so it seems to me.

2. If one event follows another according to a rule, then this is an instance of causality.

Therefore, the sun causes the stone to become warm.

According to Kant, causality is a category of the understanding. A principle of causality is a necessary condition of objective empirical knowledge. The principle of causality stipulates that every state is related to an antecedent state from which it follows according to a rule. As applied to successive states of a body in motion, the rule is given by Newton's axiom of inertial motion. All changes of a given state of motion result from forces extrinsic to the moving body. The physicist who views a change in the state of motion of a body is directed to identify an external force as cause of the change. Newton's second axiom stipulates that there is a direct proportionality between the force and the acceleration.

By linking the principle of causality to Newton's axioms of motion, Kant shifted the object of causal discourse from successive events to successive states of physical systems. Nineteenth-century physicists increasingly interpreted causality to be a regularity between states of physical systems.

Mill

Hume's official position, which became widely influential, is that a causal relation is nothing over and above the constant sequential conjunction of the members of two classes of events. John Stuart Mill noted that this position places all invariable sequences on a par. But surely some invariable sequences are noncausal. Mill pointed out that day is not the cause of night, despite their constant sequential conjunction.[25] Invariability is not a sufficient condition of causality. A causal relation must be both invariable and unconditional.[26]

Mill recommended a method of difference to uncover causal relations, namely,

instance	circumstances	phenomena
1	*ABC*	*p*
2	*BC*	—

Therefore it is probable that *A* is the cause (qua necessary condition) of *p*.

Scientists frequently apply this schema to set up controlled experiments to identify likely causes. The schema is an inductive argument. Only probability is claimed for its conclusion.

Success in applying the schema depends on the truth of the hypothesis that circumstances *A*, *B*, and *C* are the only circumstances relevant to the occurrence or nonoccurrence of *p*. If this hypothesis is true, then the method of difference does establish that *A* is the cause of *p* in the specific pair of instances under consideration. However, this success does not establish invariability. It would do so only if one could prove that what occurs in one pair of instances that instantiates the schema occurs in every pair of such instances.

Extrapolation to unexamined cases is risky. Mill called attention to the correlation between ingesting arsenic and death. Numerous instances of this correlation have been documented. However, given the presence of hydrated peroxide of iron in the subject, the correlation fails. The list of relevant circumstances must be modified such that the arsenic–death correlation is restated to hold only in the absence of this antidote.

Suppose that a particular correlation has been invariable in every case examined. There are two further requirements for causal relatedness: the invariability must extend to unexamined cases, and the correlation must be unconditional. According to Mill, a correlation is unconditional provided that it remains invariable "as long as the present constitution of things endures."[27]

To claim that the "present constitution endures" is to claim that the "ultimate laws" of the universe remain constant. Mill singled out the law of universal gravitational attraction as one such ultimate law. He contrasted the status of this law, which holds irrespective of relevant conditions, with the regular diurnal rotation of the earth that is responsible for the sequence day–night. Various events could alter the axial rotation of the earth. It is not an ultimate relation that holds in all circumstances.

Mill did not provide a list of ultimate laws that hold in all circumstances. He did, however, seek to justify causal attributions, qua invariable and unconditional relations, by appealing to the principle of the uniformity of nature.

Mill advanced an argument to support the truth of the uniformity principle. He called attention to inductive arguments of the form:

Correlation *C* is observed to hold for sample *S* of population *P*.

Therefore it is probable that correlation *C* holds for population *P*.

Mill conceded that such arguments are risky. They are at the mercy of a single exception. For example, the inductive argument that generalized from the observed color of European swans had to be abandoned with the discovery of black swans in Australia.

Nevertheless, there are numerous inductive arguments that generalize from sample to population that have proved reliable. Scientists have applied successfully generalizations about falling bodies, collisions, chemical reactions, and electromagnetic effects.

Mill conceded that arguments that proceed from what is observed of a sample to what is true of a population lack deductive force. Nevertheless, he maintained that as the scope of inductive generalization increases, so also does its reliability increase. "All swans are white" is a precarious generalization; the generalization that "for every event of a given type there is a set of circumstances on which it is invariably and unconditionally consequent" is not. Mill maintained that

> the precariousness of the method of simple enumeration is in an inverse relation to the largeness of the generalization. The process is delusive and insufficient, exactly in proportion as the subject-matter of the observation is special and limited in extent. As the sphere widens, this unscientific method becomes less and less liable to mislead and the most universal class of truths, the law of causation for instance . . . [is] duly and satisfactorily proved by that method alone.[28]

Mill maintained that every supposed exception to the principle that nature is uniform that is "sufficiently open to our observation" has been traced either to the "absence of an antecedent circumstance ordinarily present or to the presence of a circumstance ordinarily absent."[29]

Given the uniformity of nature as a premise, the method of difference is transformed from an inductive argument to a deductive argument:

1. Nature is uniform (if the pattern $\begin{matrix} A\,B\,C \text{---} p \\ B\,C \text{---} - \end{matrix}$ is true of one set of instances, it is true of every set of such instances).
2. $A\ B\ C \qquad\qquad p$
3. $\quad B\ C \qquad\qquad —$

Therefore A is the cause of p.

If the premises stating what occurs in the two instances are true, the above deductive argument explains why the conclusion is what it is. *A* is the cause of *p* because premises 1, 2, and 3 are true. According to Mill, successful causal attribution implies a sound deductive explanation of the causal relationship.

Unfortunately for Mill's position, David Hume had argued convincingly that no appeal to observations of a correlation in the past can prove that this correlation will continue to be observed in the future. Hume is correct. Mill's use of individual inductive arguments as premises in an inductive argument to prove the uniformity of nature is circular. His attempt to prove the uniformity of nature has not been judged favorably by other philosophers of science.

Notes

1. Aristotle, *Posterior Analytics*, 72^{a} 8–11.
2. Isaac Newton, *Mathematical Principles of Mathematical Philosophy*, 1686, trans. by A. Motte, revised by F. Cajori (Berkeley: University of California Press, 1962), 2:400.
3. Aristotle, *Posterior Analytics*, 89^{b} 21–26.
4. Aristotle, *Posterior Analytics*, 94^{a} 20–94^{b} 9; *Physics*, 194^{b} 24–34.
5. Aristotle, *Physics*, 198^{a} 23–27.
6. Aristotle, *Posterior Analytics*, 78^{a} 32–38.
7. Ibid., 78^{a} 29–78^{b} 2.
8. Ibid., 73^{a} 25–75^{b} 15.
9. Francis Bacon, *Novum Organum*, book 2, aphorism 11.
10. Galileo, *Bodies that Stay Atop Water or Move in It*, trans. by S. Drake, in Stillman Drake, *Cause, Experiment and Science* (Chicago: University of Chicago Press, 1981), 71.
11. Rene Descartes, *Principles of Philosophy* 1647 (Dordrecht: Reidel, 1983), 14.
12. Ibid., 12–13.
13. Ibid., 52–56; 59–63.
14. Ibid., 96; 168–69.
15. Newton, *Mathematical Principles of Natural Philosophy*, 1:393–95.
16. See, for instance, E. I. Aiton, *The Vortex Theory of Planetary Motions* (New York: American Elsevier, 1972).
17. Newton, *Opticks*, 1730 (New York: Dover, 1952), 350.
18. David Hume, *An Enquiry Concerning Human Understanding*, ed. by Stephen Buckle (Cambridge: Cambridge University Press, 2007), 59.
19. Hume, *A Treatise of Human Nature*, 1739 (Oxford: Clarendon Press, 1965), 77.
20. Hume, *An Enquiry Concerning Human Understanding*, 70.
21. Immanuel Kant, *Prolegomena to Any Future Metaphysics*, 1783 (Indianapolis: Bobbs-Merrill, 1950), 8.
22. Ibid., 58.

23. Ibid., 44.
24. Ibid., 59.
25. John Stuart Mill, *A System of Logic, 8th ed.* (London: Longman, 1970), book III, chap. 5, 222.
26. Ibid., book III, chap. 5, 221–22.
27. Ibid., book III, chap. 5, 221.
28. Ibid., book III, chap. 21, 373.
29. Ibid., book III, chap. 21, 374.

2

Regularity between States of Physical Systems in the Nineteenth Century

Scientists in the nineteenth century developed an understanding of causal relatedness that was based on emphases made by Hume and Kant. From Hume they appropriated an emphasis on regularity of sequence. Hume had taken causality to be a relation between events. Consider an instance of free fall. Hume would emphasize the regularity observed between releasing an object and its impact on the earth's surface. The former event is the cause of the latter event.

Kant shifted attention to states of physical systems and their development over time. It is states of physical systems that qualify as causes and effects. Successive states of physical systems are causally related if one state follows a second according to a rule.

What emerged from the merger of Hume's emphasis on constant sequential conjunction and Kant's emphasis on the unfolding of states of physical systems is a regularity-between-states view of causality. A nineteenth-century physicist would develop a causal analysis of free fall that describes the entire course of motion of the falling body. What is required is to select state variables for the physical system and determine how its state varies over time. Position above the earth is the obvious relevant state variable, and the appropriate functional relation is

$$\partial^2 x \,/\, \partial t^2 = g.$$

This relation, in the form of a law of odd numbers, was formulated by Galileo.[1] Galileo demonstrated that the distances of fall from rest in successive equal time intervals is the sequence 1, 3, 5, 7, 9, . . .

Figure 1
Galileo's Time-Distance Relation for Falling Bodies

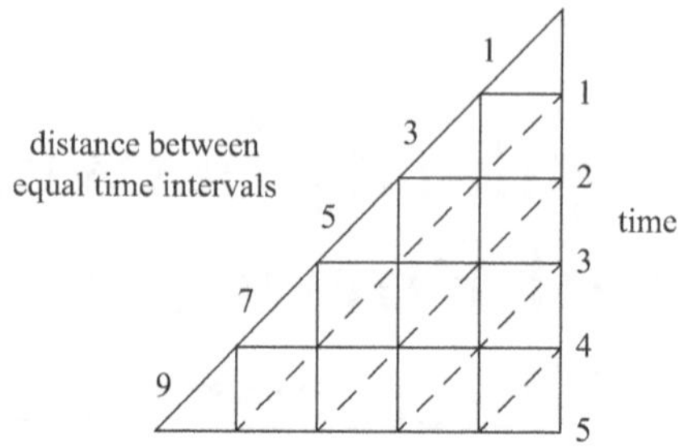

Galileo's relation holds approximately for distances of fall that are small with respect to the radius of the earth. Strictly speaking, Galileo's relation is false. Within Newtonian mechanics, force, and hence acceleration, is inversely proportional to the square of the distance between the centers of mass of two bodies. Since this distance changes as a ball falls to the earth, so too does its acceleration.

Integrating the above relation yields $x = \frac{1}{2}gt^2 + v_0t + k$. If the initial velocity is zero and the initial distance value is zero, the relation reduces to $x = \frac{1}{2}gt^2$. This is a causal relation on a regularity-between-states view of causality, since it stipulates the value of the state of the system—its position x—at successive instants of time.

F. S. C. Northrup provided a succinct statement of the regularity-between-states view of causality:

> The concept of causality as it appears in a specific scientific theory involves two factors: (1) the relation of necessary connection between the states of a system at different times, and (2) the definition of state at a given time.[2]

The regularity-between-states view is a natural modification of the regularity-between-events view championed by Hume. Hume took regularity between events to establish causal dependence (on his "official position" at least). Northrup suggested that in modern physics it is regularity between states of physical systems that establishes causal relatedness.

Bertrand Russell objected that there is no reason to label "causal" functional relations between states of physical systems. According to Russell, it suffices that physicists formulate differential equations that correlate values of variables such as position, momentum, electric field intensity, and time. These equations make no specific reference to causes and effects. Russell recommended that the term cause be restricted to the relatively underdeveloped areas of science.[3] He defended a minimalist position. Physicists utilize functional relations to describe and predict phenomena. This work can be accomplished without reference to causal relations.

Nevertheless, there is an important tradition within philosophy of science that associates functional relation, lawful regularity, and causal law. Within this tradition, causality and explanation are closely linked. The nature of this link was delineated by Karl Popper in 1934. Popper maintained that scientific explanation is achieved when causal laws are used as premises in deductive arguments, the conclusions of which are statements about specific events.[4]

On the regularity-between-states view of causality the logic of a scientific explanation takes the form

$$\begin{array}{c}(x)(Ax \supset Bx)\\ Aa\\ \hline \therefore Ba\end{array}$$

where A and B are successive states of a type of physical system, and a is a specific system of the type in question. For the argument to be explanatory, the universal conditional premise must state a causal law (and not a merely accidental correlation).

If a causal attribution—qua regularity between states—is correct, then a deductive-nomological pattern is available that explains specific changes of state. The deductive-nomological pattern stipulates the structure of a relation between causality and scientific explanation.

On the regularity-between-states view, successive states of an isolated physical system stand in the relation of cause to effect. One explains why state B was realized by citing the prior realization of state A and the generalization that all states similar to A are followed by states similar to B. The following examples illustrate the connection between the regularity-between-states theory and scientific explanation.

Regularity between States and Explanation—Examples

phenomenon	regularity	explanation
An elastic cord supporting a body of mass m is displaced distance A and released at time t_0.	The distance of the body from its equilibrium position at time t is $s = A\cos\sqrt{k/m}\,t$ and its velocity is $v = k / m(A^2 - s^2)$.	The body is at s with velocity v at time t because it was displaced at distance s and released.
A proton with velocity v enters a Wilson cloud chamber subject to magnetic field of intensity **H**.	A curved trail of water vapor is present within the chamber.	The curved path occurred because a proton of velocity v entered the chamber under a magnetic field of intensity **H**.
H_2 and I_2 are brought into contact at $t = 0$.	The concentration of HI at time t is $k[H_2][I_2]t$, given reaction $H_2 + I_2 = 2HI$.	$[HI]$ at t is $k\,[H_2]\,[I_2]\,t$ because H_2 and I_2 are brought into contact at t_0 and $d\,[HI]\,/\,dt = k$ (second-order kinetics).
A temperature gradient is established in a pure copper rod.	An amount of heat H is transmitted from B to C during time t: $H = kA(T_1 - T_2)\,t\,/\,d$, where k is the thermal conductivity of copper and A is the cross-sectional area of the rod.	The heat transfer occurred because there was established a temperature differential between points B and C.

Alternative Versions of Newtonian Mechanics

Developments in nineteenth-century physics and chemistry conformed nicely to the regularity-between-states understanding of causality. Indeed, these developments reinforced the regularity-between-states view. Among these developments were alternative axiomatizations of Newton's mechanics, classical thermodynamics, kinetic molecular theory, atomic theory, and electromagnetic theory.

Within these new physical theories, harmony prevailed with respect to causation, explanation, and prediction. To explain why state B_1 came about, one points out that state A_1 preceded it, and that because A states always are followed by B states, A_1 caused B_1. Confirmed predictions of the occurrence of further B states reinforce convictions about explanation and causal relatedness.

William Rowan Hamilton proved in 1824 that it is possible to restate Newtonian mechanics without beginning with Newton's three axioms of motion. As the starting point, Hamilton selected the state of a system of particles within which energy is conserved. For such a system, he introduced concept L, such that $L = T - V$, where T is the kinetic energy of the system and V is its kinetic energy. L is a state variable for the system of particles.

Suppose the system undergoes a transition from state A at t_0 to state B at t_1. Hamilton placed a requirement on the temporal integral of L:

> §*Ldt* is either the minimum value or the maximum value for possible transitions from state A at t_0 to state B at t_1.

If the actual path from state A to state B below is represented schematically by a solid line, then the integral §*Ldt* for this path is greater (or less) than its value for all other possible paths, represented by dotted lines.

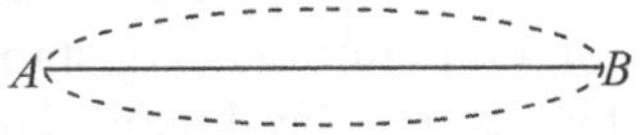

Hamilton applied the above integral principle to derive solutions to various traditional problems in mechanics, including simple harmonic motion and motion under the influence of a central force field. Whatever can be derived from Newton's axioms of motion also can be derived from Hamilton's principle.

Joseph Louis LaGrange also developed a restatement of Newtonian mechanics in terms of the time rate of change of physical systems. He did so by formulating a differential equation, solutions of which give the state of a physical system (L) at an instant of time:

$$d / dt\,(\partial L / \partial \dot{q}_j) - (\partial L / \partial q_j) = 0, j = 1, 2, 3, \ldots n$$

In this equation, $L = T - V$, and $q_1, q_2, q_3, \ldots q_n$, are generalized coordinates that represent the degrees of freedom possessed by the system, and $\dot{q}_1, \dot{q}_2, \dot{q}_3, \ldots \dot{q}_n$ are generalized velocity coordinates. If states of a system follow one another with regularity, then, on the regularity-between-states view, there is causal relatedness between these states.

Classical Thermodynamics

The development of classical thermodynamics also illustrates physicists' increasing emphasis on relations between states of physical systems. However, it is problematic whether changes of the state of a thermodynamic system count as causal. It depends on whether one insists that cause states precede effect states. The volume of a gas increases as its temperature increases (at constant pressure). However, it is not the case that the temperature increase is an event followed subsequently by a volume increase. The change in state variables is continuous. Setting aside questions about causal status, it is clear that the development of thermodynamics focused attention on physical systems and state variables.

The state of a gas sample is fully characterized by specifying the values of four variables: pressure, volume, temperature, and mass. A change in one of these variables is correlated with compensating changes in one or more of the other variables. Given a constant mass of gas, an increase of pressure at constant temperature is accompanied by a decrease of volume (Boyle's law). An increase of temperature at constant pressure is accompanied by an increase of volume (Charles's law). And an increase of temperature at constant volume is accompanied by an increase of pressure (Gay-Lussac' law).

When these laws are combined, the result is an equation of state for the system—the ideal gas law, $PV = nRT$, where n is the number of gram molecular weights of the gas and R is a universal constant for all gases.

Scientists apply the ideal gas law successfully to predict the P-V-T behavior of gases at moderate pressures and temperatures. At high pressures and temperatures, however, there is considerable divergence between calculated and observed P-V-T data. In 1873, Johannes van der Waals modified the ideal gas law to accommodate the effects of intermolecular forces and the finite volume occupied by the molecules. The resultant equation of state is

$$(P + a / V^2)(V - b) = nRT$$

where a and b are constants whose values are determined experimentally for each specific gas. The a / V^2 term takes account of intermolecular forces, and the b term takes account of the finite volume occupied by molecules of the gas. For moderate values of pressure and temperature, values calculated from the Van der Waals equation of state approach the values calculated from the ideal gas law.

Applications of the second law of thermodynamics also require specification of state variables for physical systems. One expression of the second law states that it is not possible, in a cyclic process (such as a Carnot cycle), to convert heat into work without simultaneously transferring heat from a hot to a cold reservoir.

Scientists often utilize an alternative version of the second law that refers to the entropy of a physical system. The entropy change of a thermodynamic system from state *A* to state *B* is defined as the heat absorbed by the system in a reversible process divided by the absolute temperature at which the absorption takes place—$\Delta S = (S_B - S_A) = q_{rev} / T$.

A reversible process is a path between initial and final states, each intermediate stage of which is an equilibrium state. Consider the expansion of a sample of gas. The process is reversible only in the limit of infinitely slow volume increase, such that, at every instant of the expansion, the pressure of the gas is exactly equal to the external pressure on the sample. For a reversible process at constant temperature, $P = (nR / V)$. During a rapid expansion, however, such as Joule-Thomson expansion into an evacuated container, Boyle's law cannot be applied.

In a cyclic process such as the Carnot cycle, there is a return to the initial thermodynamic state. The initial and final state variables—*P, V, T*—possess the same values.

Figure 2
The Carnot Cycle

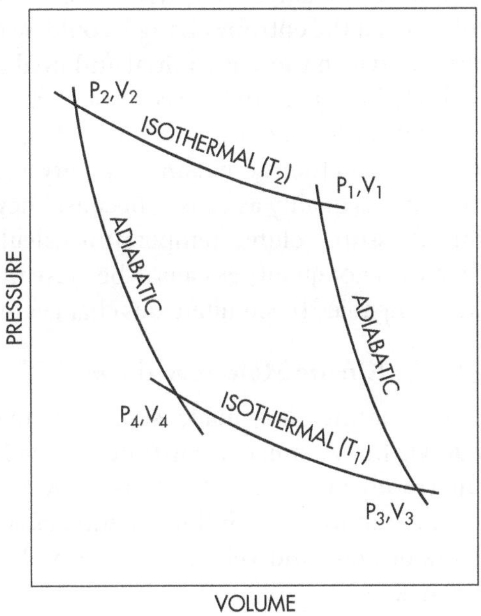

The Carnot cycle comprises four stages: (1) isothermal expansion at temperature T_2, (2) adiabatic expansion, during which there is a temperature change from T_2 to T_1, (3) isothermal compression at temperature T_1, and (4) adiabatic compression, during which there is a temperature change from T_1 to T_2. The second law of thermodynamics specifies that for a cyclic process, such as the Carnot cycle, the entropy change is zero.

$$\Delta S = \S\, dq_{rev} / T = 0$$

The entropy change for a cyclic process between states *A* and *B* of a physical system is the same regardless of the path between the states. Entropy change is a function of initial and final states alone. Max Planck observed that

> the second law . . . furnishes a relation between the quantities connected with the initial and final states of any natural process.[5]

For a noncyclic reversible process, the entropy change of a system is given by

$$\Delta S = S_B - S_A = q_{rev} / T.$$

To determine the entropy change for an irreversible noncyclic process, one must calculate what the entropy change would be for a theoretical reversible process between the same initial and final states.

If one accepts both the regularity-between-states view of causality and the requirement that cause states precede effect states, then some, but not all, changes of state in thermodynamics qualify as causal. Second-law entropy calculations qualify as causal, because they apply to states separated in time. Pressure-volume-temperature calculations based on an equation of state do not qualify as causal, because the change of one state variable is accompanied by simultaneous changes in other variables.

Kinetic Molecular Theory

Classical thermodynamics is a macroscopic theory. James Clerk Maxwell and Ludwig Boltzmann demonstrated in the 1860s and 1870s that classical thermodynamics can be derived from a theory about underlying molecular motions.[6] This kinetic molecular theory assigns the state variables of mass and velocity to each individual molecule within a sample of gas.

In the simplest version of the theory, individual molecules are treated as point particles, each of mass m. Molecules are in ceaseless motion, colliding elastically (with no loss of energy) both with one another and with the walls of the container. Given a collection of n molecules, the root-mean-square velocity μ of the collection is defined as

$$\mu = \sqrt{v_1^2 + v_2^2 + v_3^2 + \ldots v_n^2 / n}$$

The kinetic molecular theory correlates μ, a construct derived from the states of individual molecules, with the macroscopic properties pressure and temperature.

$$P = n\, m\, \mu^2 / 3\, l^3 \quad \text{and} \quad T = n\, m\, \mu^2 / 2\, k$$

Where μ is the root-mean-square velocity of the molecules,

m is the mass of each molecule,
n is the number of molecules in a cubic volume of gas, and
l is the length of one side of the cube.

The kinetic theory enables derivation of values of the macroscopic properties of a gas from state variables assigned to the microstructure via the concept of the root-mean-square velocity of the ensemble of molecules. Moreover, it is possible to deduce, from the above relations, the ideal gas law, Avogadro's principle, and Graham's law of diffusion. Avogadro's principle states that, for all gases, equal volumes at the same temperature and pressure contain an equal number of molecules. Graham's law of diffusion states that, for gases A and B, $\Delta_A / \Delta_B = \sqrt{\partial_B / \partial_A}$ where Δ is the rate of diffusion and ∂ is the density of a gas.

The kinetic theory of gases provides explanations for such macroscopic relations as the laws of Boyle, Charles, and Gay-Lussac. For example, the increase of pressure exerted by a confined gas as its temperature increases is explained by noting that the molecules of the gas move more rapidly as temperature increases, such that collisions with the confining walls of the container are both more energetic and more frequent. The kinetic theory leaves unchallenged the assumed harmony between explanation and causation. If changes of molecular velocities are believed to cause changes in the values of macroscopic properties, then citing these micro-changes is believed to explain the macro-changes.

The kinetic theory accomplished more than a deduction of previously known macroscopic laws. Maxwell deduced from the postulates of the theory that the viscosity of a gas is independent of its density at a given temperature. This was an unexpected result. Viscosity is a measure of the friction produced by layers of gas moving past one another. Intuitively, viscosity should increase with increasing density. That it does not was shown experimentally only after the deduction from kinetic theory was made.[7]

Atomic Theory

The kinetic theory of gases derives regularities at the macroscopic level from regularities assumed to hold at the microlevel. This derivation conformed to a long-standing ideal of explanation, an ideal that originated in the classical atomism of Democritus and Leucippus, and was realized in the atomic theory of the nineteenth century.

By 1800, it had become clear that there are dependable regularities in chemical reactions. The proximity of appropriate reacting species, under appropriate conditions of temperature and pressure (state 1) invariably are followed by the presence of product species (state 2). There is causal relatedness *qua* regularity between states.

Supporters of the atomic theory maintained that there exist regularities at a microlevel that are associated with the causal sequences at the macrolevel. Moreover, they held that these microlevel interactions *cause* the regularities between states observed at the macrolevel. This position requires a commitment to the existence of micro-entities.

Consider the formation of salt. Proximity of sodium and chlorine invariably is followed by the presence of sodium chloride. To produce salt, it takes 35.5 grams of chlorine gas to convert 23 grams of sodium into salt. If the reaction is accomplished instead, on 2.3 grams of sodium, or 0.023 grams of sodium, the weight ratio of sodium to chlorine remains 23 / 35.5. Reactions between chemical species are subject to a law of definite proportions.

Since weight ratios remain constant as the weights of reacting substances are decreased, it seems plausible to hypothesize that there exist minimum portions of chemical species. Of course, it is logically possible that substances are infinitely divisible. However, on the atomic theory, when sodium reacts with chlorine, large numbers of sodium atoms combine one by one with chlorine atoms.

John Dalton (1766–1844) was largely responsible for the successful revival of the classical atomic theory of matter. He made the law of definite proportions a basic principle of atomic theory. He noted, in addition, that when two chemical species combine to form a series of different compounds, the weights of the two species are in small whole number ratios. Nitrogen and oxygen provide a good illustration. These two chemical species form separate compounds with weight ratios 2 : 1, 1 : 1, 1 : 2, 2 : 3, and 2 : 5. Dalton insisted that the best explanation of this law of multiple proportions is that individual atoms of nitrogen and oxygen combine with one another in precisely these ratios.

In order to represent chemical reactions as combinations of atoms, Dalton affirmed a "principle of greatest simplicity."[8] This principle requires the theorist to assign the formula *AB* to the single compound formed by *A* and *B*, the formulas *CD* and either C_2D or CD_2 to the two compounds formed by *C* and *D*, and so on. Application of this principle led to assigning the formula HO to water (believed at the time to be the only compound of hydrogen and oxygen). This assignment, given the known 1 : 8 weight ratio in water, assigned an atomic weight of 8 to oxygen, relative to hydrogen at 1. This assignment was puzzling. Clearly water (HO) must be heavier than oxygen (O), but it was known that the density of water vapor is less than the density of oxygen. A further difficulty for Dalton was the volume relation in the reaction that produces carbon monoxide

$$\text{C} + \text{oxygen (1 volume)} = \text{CO (2 volumes)}.$$

It seemed to Dalton that equal volumes of gases may contain unequal numbers of particles. In the above reaction, there must be twice as many oxygen particles as carbon monoxide particles in a given volume of gas.

Amadeo Avogadro indicated a way out of these difficulties in 1810.[9] Avogadro maintained that equal volumes of gases contain equal numbers of particles, but that in many cases the particles of a gaseous element are molecules composed of two atoms. Given that gaseous oxygen occurs as O_2, the above weight and volume relations no longer are puzzling. Unfortunately chemists at that time did not accept the hypothesis of polyatomic molecules. Studies by Davy and Berzelius had shown that, in many cases, atoms combine to form molecules because of electrostatic attraction. For example, electropositive sodium and electronegative chlorine exhibit an elective affinity for one another because of electrostatic attraction. It was difficult to see how there could be a force of attraction that would hold together two atoms of the same element. Avogadro's hypothesis was restated by Stanislao Cannizzaro in 1860, and it was accepted by most chemists at that time.

Commitment to the existence of atoms was just one of a number of competing research programs for chemistry in the mid-nineteenth century. Other influential programs included: (1) a positivistic rejection of theoretical entities, (2) mathematical Newtonianism, and 3) a pursuit of unity-of-matter theories.

The positivist position was to remain agnostic on the existence of atoms. Davy and Whewell stressed that the empirical content of atomic theory is the laws of definite proportions and multiple proportions. References to interactions of atoms are superfluous. The theory of tiny invisible atoms associating and dissociating is a mere "picture preference," commitment to which adds nothing to progress in science.

Mathematical Newtonianism is the position that chemical reactions are the results of central-force interactions between particles. Atomic theory lacks mathematical sophistication. LaPlace and Berthollet called upon chemists to show how chemical reactions obey short-range force laws—$\mathbf{F} \propto 1 / r^n$, with $n > 2$—modeled on Newton's law of gravitational attraction.

Unity-of-matter theorists insisted on the identity of the ultimate constituents of matter. They rejected the atomists' belief that there exist multiple distinct chemical atoms that differ with respect to weight. William Prout suggested in 1810 that hydrogen is the "prime matter" from which all elements are formed.[10] Some early atomic weight determinations were consistent with Prout's "protyle hypothesis" (e.g., nitrogen = 14, sulfur = 32, and phosphorus = 40). Most chemists lost interest in the unity-of-matter approach when other atomic weight determinations failed to yield an integral result (e.g., chlorine = 35.5).

Imre Lakatos cited Prout's protyle hypothesis as an example of a failed research program that subsequently was revived.[11] In the twentieth century, the discovery that most elements occur in nature as mixtures of isotopes enabled the Proutian research program to be reinstated for atomic numbers (the number of protons in the nucleus) rather than atomic weights.

Benjamin Brodie appealed to each of the three strands of opposition to the atomic theory in his 1869 presidential address to the Chemical Society. He had become increasingly disenchanted with the atomic theory in the course of his research. Brodie reported distaste at viewing an advertisement of a set of balls and rods for the construction of model molecules. In 1866 he published an alternative to Dalton's theory.[12] He called it a "calculus of chemical operations." The basic principles of the calculus are:

1. Greek letters stand for operations performed on a unit of space (one liter at $T = 25°$ centigrade and $P = 1.0$ atm.), such that changes in weight occur, for example,
   α—a unit of space is converted into a unit of hydrogen.
2. α, the hydrogenation of one unit of space, is the basic operation that is undistributed in all combinations. Other operations include
   ξ—a unit of hydrogenated space is converted into a unit of water, and
   χ—a unit of hydrogenated space is converted into a unit of hydrogen chloride.
3. $\alpha\omega = \alpha + \omega$, where $\alpha\omega$ is the result of successive operations of α and ω on a unit of space.

Applied to the decomposition of hydrogen chloride, the calculus yields the formula

$$2\,\alpha\,\chi = \alpha + \alpha\,\chi^2, \text{ namely,}$$

2 vol. hydrogen chloride = 1 vol. hydrogen + 1 vol. chlorine

$$2\,\alpha^m\,\chi^{m'} = \alpha + \alpha^n\,\chi^{n'}$$

$$\alpha^m\,\chi^{m'} + \alpha^m\,\chi^{m'} = \alpha + \alpha^n\,\chi^{n'}$$

$$(\alpha^m\,\chi^{m'})^2 = \alpha\,\alpha^n\,\chi^{n'}$$

$$\text{thus } 2m = (1 + n), \text{ and } 2m' = n'$$

assume the simplest integral solutions

$$m = 1,\ m' = 1,\ n = 1,\ n' = 2$$

$$\text{or, } \quad 2\,\alpha\,\chi \quad = \quad \alpha + \quad \alpha\,\chi^2$$

hydrogen chloride hydrogen chlorine

This is an interesting result. It suggests that chlorine may be a compound of hydrogen and a yet-to-be-discovered element whose operation is χ.

Not surprisingly, Brodie abandoned the vantage point of positivism in his defense of his calculus. He observed that the formulas of certain elements turn out to be complex, for instance, chlorine, $\alpha\chi^2$; bromine, $\alpha\beta^2$; nitrogen, $\alpha\, \upsilon^2$; and phosphorus, $\alpha^2\varphi^4$. Brodie suggested that there may exist yet-to-be-discovered elements corresponding to such operations as χ, β, υ, and φ.[13] He noted that William Huggins had obtained spectra from nebulae that contain emission lines that cannot be attributed to known terrestrial elements. The lines were attributed to Nebulium, a substance unknown on earth. These lines actually are produced by forbidden transitions of doubly ionized oxygen (O^{++}) which occur under conditions of low pressure (high mean-free path).

Fertility is one criterion of acceptability for scientific theories, and Brodie's Calculus did stimulate a modest amount of research. In particular, it provided a rationale for the search for new elements in spectra from stars and nebulae. The search was at least partly successful. Norman Lockyear, a former research associate of Brodie, discovered helium in the sun's atmosphere in 1870. Other investigators had noticed the helium emission lines in the solar spectrum, but they uniformly had attributed the lines to hydrogen or some other element also present in the terrestrial environment. Terrestrial helium was identified only in 1895 by William Ramsey.

Brodie's chemical calculus aroused a fair amount of interest. In 1867, a joint meeting of the Chemical Society and the Royal Society was convened to debate its merits. Criticisms of the calculus were advanced at this meeting, and subsequently, for which Brodie had no reply.

W. S. Jevons objected to the basic formula ($xy = x + y$). He noted that the formula is inconsistent with the basic distributive law of algebra: $x\,(y + z) = xy + xz$.[14]

Friedrich Kekulé observed that Brodie had taken the hydrogenation of one unit of space to be the basic operation, and that, on this assumption, the symbol for chlorine ($\alpha\chi^2$) is complex. Kekulé pointed out that if chlorination of one unit of space is taken to be the basic operation, then hydrogen is assigned a complex symbol. According to Kekulé, Brodie's calculus is unacceptably arbitrary.[15]

In addition, it was shown that Brodie's calculus leads to results that conflict with empirical evidence. Edward Frankland maintained that, within Brodie's calculus, sulfuric acid is ($\alpha\,\theta\,\xi^4$) and nitric acid is ($\alpha\,\upsilon\,\xi^3$). This suggests that the hydrogen in these two acids can be removed in just one stage. However, it was well established that the hydrogen in sulfuric acid (H_2SO_4) must be removed in two stages, whereas the hydrogen in nitric acid (HNO_3) can be removed in only one stage. The formulas from Dalton's atomic theory are more accurate.[16]

Brodie's chemical calculus became a degenerating research program (in Lakatos's sense) during the last decades of the nineteenth century. It failed to generate enough interesting new results, and it was eclipsed by an atomic theory newly made consistent by Cannizzaro's revival of Avogadro's hypothesis of polyatomic molecules.

The atomic theory received support from the success of kinetic molecular theory. This was an important development, since it provided support from physics for what had been predominantly a chemical theory. Kinetic molecular theory achieved a correlation between a mathematical analysis of elastic collisions between minute particles and the walls of their containers, and changes in the macroscopic properties of gases. Most scientists concluded that, since there is extensive evidence that momentum and energy are conserved in elastic collisions between macroscopic bodies, collisions between molecules qualify as genuine causes of the observed behavior of gases.

Additional support was provided by Van't Hoff and Le Bel in 1874.[17] These investigators successfully applied the atomic theory to cases of geometrical isomerism, for example,

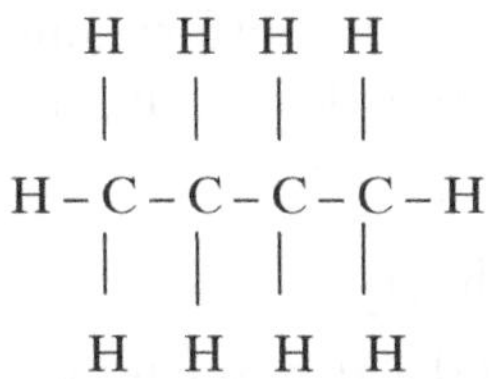

butane

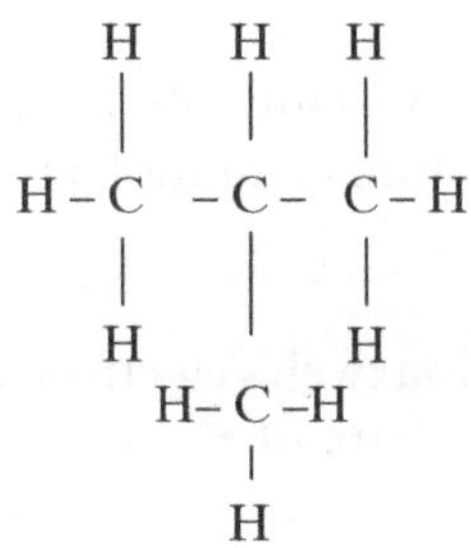

isobutane

and optical isomerism, for example

(dextrorotatory and levorotatory lactic acid)

Ernest Rutherford (1909) and Jean Perrin (1912) provided even more decisive support for the existence of atoms and molecules. Rutherford and coworkers bombarded various elements with α-rays and observed that most α-particles passed through with minimum deflection, but that some α-particles were deflected through angles greater than 90 degrees.[18] The obvious conclusion was that elements, such as gold and aluminum, have a microstructure of very dense atoms, each of which is surrounded by a large area of empty space.

Perrin called attention to the incredible agreement of determinations of the value of Avogadro's number (N_o) from diverse experimental arrangements.[19] Avogadro's number is the number of molecules in a gram molecular weight of an element or compound. On the kinetic theory, this number is the number of molecules in 22.4 liters of any gas under standard conditions—25 degrees centigrade and 1 atmosphere of pressure. N_0 has been calculated from the equations of electrolytic deposition, radioactive decay, Brownian motion, and black-body radiation. These equations govern very different physical processes. Nevertheless, values calculated for N_0 from these equations converge on the value 6.02×10^{23} molecules/gram molecular weight. After the work of Rutherford and Perrin, it was an untenable position to remain agnostic on the existence of atoms and molecules.

Electromagnetic Theory

Maxwell's electrodynamic theory is one of the great scientific achievements of the nineteenth century. The theory is encapsulated in a set of equations that correlate state variables, such as electric field intensity (**E**) and magnetic field intensity (**H**).

There is a problem, however, about how to characterize the state of an electrodynamic field. State variables **E** and **H** state potentials that become actualized only in the presence of some body at the point under consideration.

Consider the famous experiment of Oersted.[20] Oersted placed a magnetized needle in the air above a current-carrying wire. The needle assumed an orientation at a right angle to the wire. When Oersted reversed the direction of current flow in the wire, the needle pivoted through an angle of 180 degrees. Moreover, for a given direction of current flow, a needle placed beneath the wire took up an orientation 180 degrees from the orientation of a needle placed above the wire.

Figure 3
Oersted's Experiment[21]

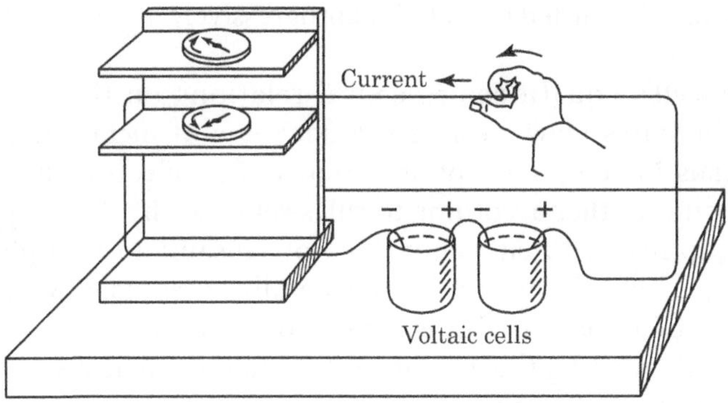

Michael Faraday hypothesized that there were concentric lines of magnetic force around the current-carrying wire. These lines of force presumably exist whether or not there are objects present in the field. They exist as a field of potential effects whose magnitude decreases with increasing distance from the wire. These potential effects can be cashed by appropriate operations of measurement. If a unit magnetic pole were placed at a distance x from the wire, then it would experience a specific force in a specific direction.

This is all well and good. But how is one to define the state of the system when the system is an electrodynamic field? At a particular time one can test for a state variable at a specific location in space, but the state variable also has values at that time in other regions of space. A field is not a system localized in space.

F. C. S. Northrup pointed out that it is necessary to introduce "retarded potentials" in order to complete the state description for an electrodynamic system. He declared that

> only at the point in the field where the observer is located at the present time t_0 do non-retarded potentials, (i.e., directions and intensities of the forces at the time t_0 occur). At all distant points the potentials which enter into the definition of state at the present time t_0 are directions and intensities of the force at those points, not at the time t_0 but at an earlier time t_n such that these forces, leaving the distant points at the time t_n and traveling with the finite velocity c, arrive at the observer's position at the present time t_0.[22]

In this respect, state description for electromagnetic theory differs from state description for Newtonian mechanics and for statistical mechanics. The state of an electrodynamic system includes both values of electric and magnetic potentials at the point under observation and values of retarded potentials at progressively increasing distances from that point.

Maxwell's equations are causal relations on the regularity-between-states view, despite the necessity of including retarded potentials in state descriptions. States of an electrodynamic field follow one another according to rules set down by these equations. Northrup insisted that "the validity of causality in field physics as well as particle physics has been generally accepted."[23] Moreover, if a change of a state variable is judged to be the cause of a particular motion, then citing this change of the state variable also counts as an explanation of that motion.

The successes achieved by the theories discussed above promoted confidence among scientists that the discovery of causal relations —qua regularities between states of physical systems—leads to appropriate explanations of physical processes. This confidence that prevailed among scientists at the end of the nineteenth century was to be tested severely by developments in the early twentieth century.

Notes

1. Galileo, *Two New Sciences*, 1638, trans. by S. Drake (Madison: University of Wisconsin Press, 1974), 167.
2. F. S. C. Northrup, "Causality in Field Physics in its Bearing upon Biological Causation," *Phil. Sci.* 5 (1938), 166.
3. Bertrand Russell, "On the Notion of Cause," 1912, in *Mysticism and Logic* (New York: Doubleday Anchor, 1917), 181.
4. Karl Popper, *The Logic of Scientific Discovery* (New York: Basic Books, 1959), 59–64.
5. Max Planck, *Treatise on Thermodynamics*, 1922, (New York: Dover), 87.
6. See, for instance, *Kinetic Theory*, 2 vols., ed. by Stephen G. Brush (Oxford: Pergamon Press, 1948).
7. James Clerk Maxwell, "Illustrations of the Dynamical Theory of Gases," *Phil. Mag.*(1859) 19–26. Reprinted in *Kinetic Theory*, 1:148–49.
8. John Dalton, "On Chemical Synthesis," in *A New System of Chemical Philosophy*, 1808 (New York: The Citadel Press, 1964), 163–65.
9. Amadeo Avogadro, "Essay on a Manner of Determining the Relative Masses of the Elementary Molecules of Bodies . . .," 1811, in *Foundations of Molecular Theory: Alembic Club Reprint No. 4* (Edinburgh: The Alembic Club, 1969), 28–51.
10. William Prout, "On the Relation between the Specific Gravities of Bodies in their Gaseous State and the Weights of their Atoms," 1815, in *Prout's Hypothesis: Alembic Club Reprint No. 20* (Edinburgh: The Alembic Club, 1969), 25–40.

11. Imre Lakatos, "Falsification and the Methodology of Scientific Research Programmes," in *Criticism and the Growth of Knowledge*, ed. by I. Lakatos and A. Musgrave (Cambridge: Cambridge University Press, 1970), 138–40.
12. Benjamin C. Brodie, "Calculus of Chemical Operations," *Phil. Trans. Royal Soc. 156* (1866), 781–859.
13. W. H. Brock and D. M. Knight, "The Atomic Debates," in *The Atomic Debates*, ed. by W. H. Brock (Leicester: Leicester University Press, 1967), 13–15.
14. D. M. Dallas, "The Chemical Calculus of Benjamin Brodie," in *The Atomic Debates*, ed. by W. H. Brock (Leicester: Leicester University Press, 1967), 55–56.
15. Ibid., 57–58.
16. Ibid., 49.
17. See, for instance, Aaron J. Ihde, *The Development of Modern Chemistry* (New York: Harper & Row, 1964) 324–9; 621–31.
18. Ernest Rutherford, "Forty Years of Physics," in *Background to Modern Science*, ed. by J. Needham and W. Pagel (New York: MacMillan, 1938), 68–69.
19. Jean Perrin, *Atoms*, trans. by D. Hammick (London: Constable, 1922), 215–17.
20. H. C. Oersted, "Experiments on the Effect of a Current of Electricity on the Magnetic Needle," 1820, in *Source Book in Physics*, ed. by W. F. Magee (Cambridge: Harvard University Press, 1965), 438–39.
21. Guy Omer, Jr., Harold L. Knowles, Beverly W. Mundy, and Herbert Yoho, *Physical Science: Men and Concepts* (Boston: D. C. Heath, 1968), 530.
22. Northrup, "Causality in Field Physics . . .," 171.
23. Ibid., 170.

3

Prelude to Quantum Mechanics

Radioactive Decay

In the early years of the twentieth century, Ernest Rutherford undertook pioneering studies of radioactive decay sequences. The discovery of radioactivity provided support for C. S. Peirce's claim that chance is an objective feature of the universe.[1]

The decay of an atom of a radioactive element is spontaneous. On the regularity-between-states theory of causality, there is no identifiable antecedent state from which the emission state regularly follows. It would seem that, if any event is uncaused, it is the emission of an α-particle.

The most that can be said of an individual emission event is that it had a certain probability of occurrence. Of course, radioactive species have half-lives. Given a very large number of atoms of a radioactive species at time t, one-half will emit particles by time $t + \Delta t$, where Δt is the half-life of the species. Which particular atoms will decay during this period cannot be determined in advance.

Kinetic molecular theory and the theory of radioactive decay both are based on statistical considerations. In the kinetic theory, molecules of a gas are assumed to have a statistical distribution of velocities. It is possible, in principle, to ascertain the state of an individual molecule. In radioactive decay, by contrast, emission of an α-particle from a nucleus is more or less probable, depending on the half-life of the species. However, there is no prospect of determining the instantaneous state of a nucleus of a radioactive species.

Is the existence of radioactivity a threat to the "causality implies explanation" thesis? It clearly is not a threat at the level of individual emission events, since no claim is made that the decay of an atom is the effect of some antecedent cause. However, there is an observed regularity of states during the decay of a sample of the element. Rutherford published a decay curve for Thorium in 1906:

Figure 4
Rutherford's Decay of Activity of the Thorium Emanation[2]

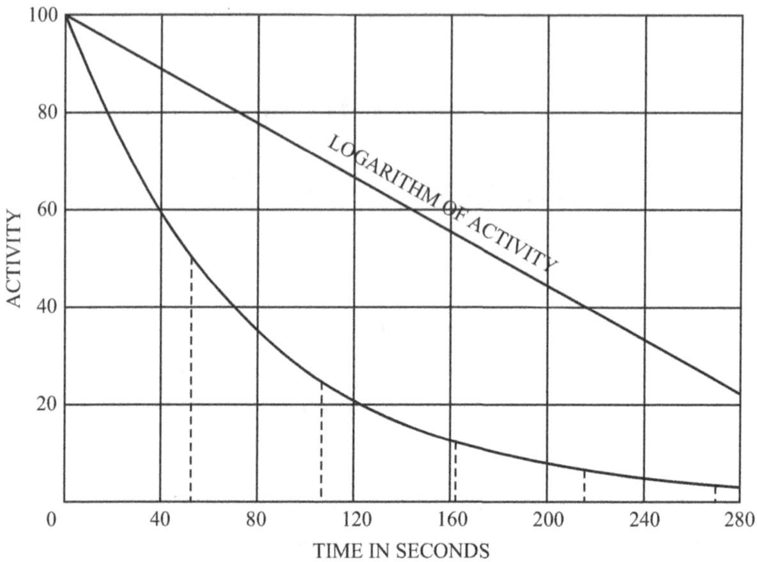

On the regularity-between-states view of causality, each stage of the decay process is in a cause-effect relation to its successor stage. The stages thus related are states of the entire ensemble of atoms and not the states of individual atoms. Nevertheless, this presents a problem for the "causality implies explanation" thesis. The decay of thorium, qua ensemble of atoms, qualifies as a causal process on the regularity-between-states view.

Does knowledge that successive stages of radioactive decay are causally related have explanatory value? One might argue that it does, because radioactive decay processes, like cases of inertial motion, are exceptional cases for which regularity of sequence is self-explanatory. Uniform rectilinear motion is self-explanatory, or, rather, inertial motion (motion of a body in the absence of all external forces) *would be* self-explanatory if it could be realized. The exponential decrease of radioactive intensity also would seem to be self-explanatory. Of course, there is an important difference between the two cases. In the case of inertial motion, there is an entity that persists throughout the process. In the case of radioactive decay, a nucleus is transformed into a daughter nucleus (or nuclei), subatomic particles, and energy.

Knowledge about radioactive decay processes does not disprove the "causality implies explanation" thesis. Nevertheless, the claim that exponential decay patterns are self-explanatory does not rise to the level of those causal explanations that cite changes in momentum or energy. Given what is known about radioactivity, it is reasonable to doubt that causal attribution invariably counts as an explanation.

Thomson and Rutherford on Atomic Structure

It became clear from the work of Becquerel, Madame Curie, Rutherford, and others that certain atoms are unstable. In 1906 Rutherford could offer only a vague picture of the internal structure of the atom. He wrote that

> the atom is supposed to consist of a number of charged parts in rapid oscillatory or orbital motion and consequently contains a great store of energy. Part of this energy is kinetic and part potential, resulting from the condensation of the electrical charges within the minute volume of the atom. This latent energy of the atom does not ordinarily manifest itself. . . . Part of this energy is, however, released in radioactive changes, where the atom itself suffers disruption with the expulsion of one of its charged parts with great velocity.[3]

J. J. Thomson sought to reduce the vagueness. He advanced a picture of atomic structure that distributes negatively charged spheres within a larger sphere of positive charge. Thomson's picture became known as the "plum pudding" model of the atom. For each element, Thomson assigned that number of embedded negative spheres that corresponds to its position in the Periodic Table. Two-dimensional representations of helium and oxygen are given below:

Figure 5
The Plum Pudding Model

Helium (2 negative spheres) — Oxygen (8 negative spheres)

The "plum pudding" model quickly was shown to be untenable by experiments performed in Rutherford's Manchester laboratory.

Hans Geiger passed a beam of α-particles through a sheet of gold foil and found that there were minor deflections produced by the motion of the particles through the foil. Rutherford then suggested that a search be undertaken for any large deflections. Ernest Marsden conducted a search and found that the occasional α-particle was deflected through an angle greater than 90 degrees. This result was inconsistent with the "plum pudding" model. Rutherford and his associates demonstrated that the amount of large-angle deflection increases with increasing atomic weight of atoms in the foil.

Rutherford took the results of the Geiger-Marsden experiments to require a nuclear model of the atom. He maintained that

> the positive charge associated with an atom is concentrated into a minute centre or nucleus and that the compensating negative charge is distributed over a sphere of radius comparable to the radius of the atom.[4]

Rutherford maintained that the extreme deflection of α-particles results from a Coulomb's law repulsion between the particle and a positively charged nucleus:

$$F = k\, q_\alpha\, q_n \,/\, r^2.$$

He noted that scattering results show that for angles greater than one degree, the Coulomb's law attractive force between α-particles and negative charges in the atom may be neglected.[5]

Given Coulomb's law and evidence from scattering experiments, a rough estimate can be made of the dimensions of various nuclei. Rutherford concluded that the maximum value for the radius of a gold nucleus is 3.2×10^{-12} cm. Since the radius of the gold atom is about 1×10^{-8} cm, it is clear that nuclear dimensions are minute compared to atomic dimensions.[6]

Radioactive decay involves the ejection of α-particles, electrons, and γ-rays from the nucleus of an atom. Absorption spectra and emission spectra indicate that discrete energy transitions also take place within atoms without the ejection of particles.

Absorption spectra were the first to be studied. Discrete lines were observed in the spectrum of the sun by Wollaston and Frauenhofer in the early years of the nineteenth century. These lines are produced by absorption of energy at specific wavelengths by elements in the sun's atmosphere.

Emission spectra were obtained in European laboratories in the mid-nineteenth century. Angstrom, Kirchoff, Bunsen, Crookes, and others established that elements in gaseous form, heated to high temperatures at low densities, produce spectra containing bright lines at specific wavelengths. For a given element, emission lines and absorption lines occur at the same wavelengths.

The spectrum of hydrogen was of particular interest. Hydrogen, after all, is the simplest of the elements. Johann Balmer pointed out in 1885 that there is an interesting mathematical relationship governing the positions of lines of the hydrogen spectrum. The wavelengths of these lines are given by the formula

$$\lambda = kn^2 / n^2 - 2^2$$

where k is a constant and n is integer 3 or greater. This collection of spectral lines, with n = 3, 4, 5, . . . is referred to as the Balmer series.

Figure 6
The Balmer Series Spectral Lines

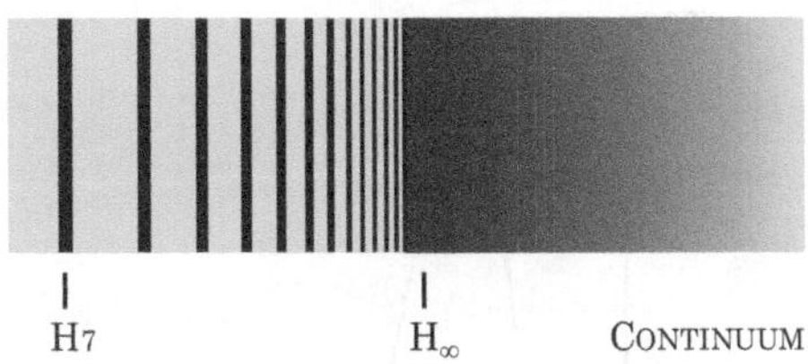

Scientists soon discovered additional series of spectral lines in the hydrogen spectrum. These series conform to the formulas: $\lambda = kn^2 / n^2 - 1^2$ (Lyman); $\lambda = kn^2 / n^2 - 3^2$ (Paschen); $\lambda = kn^2 / n^2 - 4^2$ (Brackett); and $\lambda = kn^2 / n^2 - 5^2$ (Pfund).

This situation was puzzling. Regularities had been established that link the heating of elements and the occurrence of lines in their spectra. Moreover, lines in the hydrogen spectrum are grouped in five series that obey simple numerical relationships. Hydrogen atoms were known to possess just one negative charge and one positive charge. On Rutherford's nuclear model, for which there was extensive evidential support, the single electron somehow travels around its nucleus, a single proton. When the atom acquires energy upon heating, it releases this energy in discrete packets, always at specific wavelengths. The resources of Newtonian mechanics and electrodynamic theory appeared to be inadequate to account for this spectral evidence.

Planck's Quantum Hypothesis

One key to the solution of the puzzle was a hypothesis proposed by Max Planck. Planck had been working on the problem of black-body radiation, a phenomenon whose relation to spectral emission was not obvious at the time. Black-body radiation, unlike emission spectra, is radiation that appears to be continuous over a wide range of wavelengths.

A black box is a box with black interior walls, one of which contains a small hole. To produce black-body radiation the box is heated to, and maintained at, a specific temperature. One then may measure the radiant energy emerging from the hole. The observed distribution of energy over a range of wavelengths is displayed below:

Figure 7
Black-Body Radiation

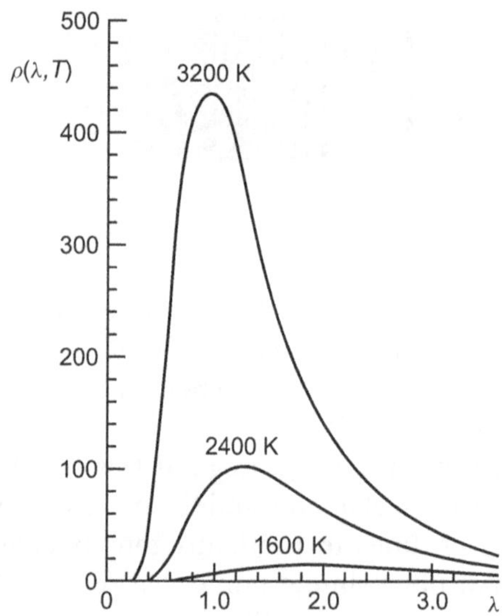

A number of formulas were proposed in an attempt to reproduce the observed energy-wavelength curves. Wilhelm Wien (1896) proposed the formula

$$E_\lambda = a \,/\, \lambda^5 \exp^{b/\lambda T},$$

where E_λ is the radiant energy at wavelength λ, T is temperature, and a and b are constants. Wien's formula fits the energy-wavelength curves at short wavelengths. It is not accurate at longer wavelengths, however.

Lord Rayleigh and James Jeans (1900) proposed the formula

$$E_\lambda = 2\pi kT \,/\, c\lambda^4,$$

where k is Boltzmann's constant and c is the velocity of light. The Rayleigh-Jeans formula fits the energy-wavelength curves at long wavelengths. It is a disaster at short wavelengths, however, since it predicts that radiant energy increases without limit as wavelengths decrease. This consequence of the Rayleigh-Jeans formula became known as the "ultraviolet catastrophe."

Max Planck (1900) produced a formula that succeeded in reproducing the observed energy-wavelength curves at both long and short wavelengths. He hypothesized that there exist a large number of linear harmonic oscillators within the black-body cavity. In order to establish agreement with observed energy-wavelength curves Planck restricted the possible energy values of an oscillator to integral multiples of $h\,v$, where v is the frequency of oscillation and h is a constant whose value is 6.62×10^{-27} erg-sec. Energy emitted by a black-body thus is quantized. Planck's formula is

$$E_\lambda = 2\pi hc^2 \,/\, \lambda^5 \exp^{(ch/\lambda kT-1)}.$$

Planck's formula was a success. At short wavelengths it yields values that coincide with those calculated from Wien's formula. At long wavelengths Planck's equation yields values that coincide with those calculated from the Rayleigh-Jeans formula. Max Jammer observed that

> Planck's introduction of h seems to have been regarded at that time as an expedient methodological device of no deeper significance, although his radiation law was repeatedly subjected to experimental test.[7]

The hypothesis of energy quantization became more than "an expedient methodological device" when Einstein showed that energy values in the photoelectric effect also are quantized. In one respect, Einstein's position is an extension of Planck's hypothesis. Whereas

Planck suggested that black-body oscillators responsible for radiation exist in quantized energy states, Einstein maintained that the emitted light itself exists in quantized bundles or photons. In the photoelectric effect, electrons are set free on the surface of a metal upon impact of a beam of light. Einstein's account of this process is depicted below:

Figure 8
Einstein's Account of the Photoelectric Effect

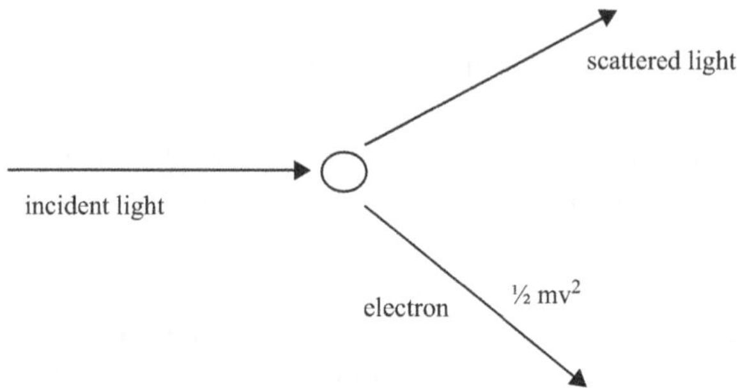

Bohr's Early Theory of the Hydrogen Atom

Niels Bohr wrote a PhD dissertation in 1911 at the University of Copenhagen on the electron theory of metals. J. J. Thomson, and others, had attributed electrical conductivity in metals to the motion of free electrons. Bohr's paper dealt with the relation between the motions of electrons and measured values of thermal and electrical conductivities.

Upon receipt of his degree, Bohr left Copenhagen for Cambridge to work with Thomson on the electron theory of metals. The collaborative effort envisioned by Bohr did not materialize, however, and Bohr elected in 1912 to join Ernest Rutherford's group at the University of Manchester. Rutherford had suggested a nuclear model of the atom, and Bohr developed nuclear models that specified the arrangement of electrons in atoms of various elements.

Bohr returned to Copenhagen in 1913, at which time he became aware of the mathematical relation that specifies the positions of the lines of the Balmer series in the hydrogen spectrum. Bohr recognized that the Balmer series could be explained as a series of transitions within the hydrogen atom from one energy level to another, on the assumption that these energy levels correspond to discrete orbits of the electron.

Bohr made the quantization of energy a cornerstone of his theory of the hydrogen atom (1913). Henry Folse emphasized that Bohr was greatly impressed by the considerable evidence for the stability of the atom throughout chemical reactions.[8] Bohr sought to account for energy transitions within the atom without undermining its fundamental stability. He assigned a set of discrete orbits to the hydrogen electron, and attributed the emission of energy to transitions between these orbits. By so doing, he was able to correlate transitions within the atom with observed spectral lines. Bohr's theory is based on four principles:

1. The electron is constrained to move along one of a collection of circular orbits centered on the proton.
2. A quantum condition applies. The angular momentum of the electron in an orbit is given by $mv = nh / 2\pi r$, where n is an integer, and h is Planck's constant.
3. In its circular motion around the nucleus, the electron obeys both Newton's second axiom ($\mathbf{F} = m\mathbf{a}$) and Coulomb's law of electrostatic attraction ($\boldsymbol{F} = k\, q_p\, q_e / r^2$), where q_p is the charge of the proton, q_e is the charge of the electron, and k is a constant.
4. A restriction applies on the absorption and emission of energy. No energy is absorbed or emitted by an electron as it circles in its orbit. Transitions from one electron orbit to another are accompanied by an energy change whose magnitude is given by $\Delta E = (n_f - n_i)\, hc / \lambda$, where n_f and n_i are the integers representing the final and initial orbits. Orbits are numbered 1, 2, 3, . . . in order of increasing radius.

By correlating values of n with values of wavelength λ, Bohr was able to show that transitions from $n = 3, 4, 5 \ldots$ to $n = 2$ correspond to lines in the Balmer series, and that transitions to $n = 1$, $n = 3$ and $n = 4$ correspond to lines in the Lyman, Paschen, and Brackett series.

In 1896, F. C. Pickering had found a series of spectral lines in the star Zeta Puppis. Pickering, Rydberg, and other investigators took these lines to be part of the hydrogen spectrum under extreme conditions of temperature and pressure. The Pickering lines fit the formula for the Balmer series provided that half-integral numbers were allowed. These lines conform to the formula

$$\upsilon = [1 / 2^2 - 1 / \mathrm{n}^2].$$

John Heilbron emphasized that Bohr refused to accept half-integral transitions.[10] Such transitions would violate the quantum condition on which his theory of the hydrogen atom is based. Bohr rewrote the formula as

$$\upsilon = 4R\, [1 / 4^2 - 1 / n^2],$$

Figure 9
Bohr on Electronic Transitions in the Hydrogen Atom[9]

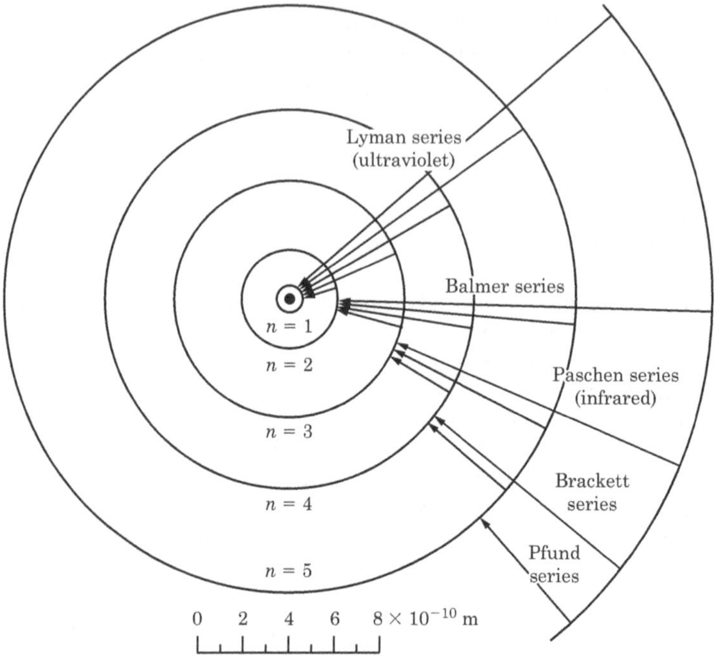

and attributed the series to electron transitions in singly ionized helium, which has lost one of its two orbital electrons.

Bohr took the general frequency formula for electron transitions in an atom to be

$$v = RZ^2\,[1 / n_2^{\,2} - 1 / n_1^{\,2}],$$

where Z is the charge of the nucleus, and $R = 2\pi^2 me^4 / ch^3$. Given $Z = 2$ for helium, he concluded that RZ^2 for helium should be four times the value for hydrogen.

Bohr's theory achieved predictive success in this instance. Spectroscopists subsequently produced the lines of the Pickering series from specimens of purified helium gas. Values calculated from Bohr's formula were in rough agreement with the positions of the observed spectral lines. Bohr then improved the initial rough agreement by taking account of the effect on the orbital electron's motion of the finite masses of the hydrogen nucleus and the electron. He replaced the term m in the above equation for R by $m^* = m_e(1 + m_e / m_{He})$, thereby improving agreement with the spectroscopic data. The predictive success thus achieved was an important point in favor of the Bohr theory.

Nevertheless, the Bohr theory achieved only limited success. There were features of the observed hydrogen spectrum that did not fit calculations from the theory.

Spectroscopists reported in the early years of the twentieth century that the lines of the Balmer series were not sharp single lines. Rather, each of these dominant lines in the hydrogen spectrum was, in reality, two or more separate, closely spaced, lines. Arnold Sommerfeld declared in 1916 that Bohr's theory of the hydrogen atom provided only a first approximation to its structure, since it did not account for the splitting of spectral lines.[11]

Sommerfeld extended the Bohr theory to incorporate transitions between elliptical electron orbits. By so doing, he was able to explain many of the puzzling features of the hydrogen spectrum.

Figure 10
Sommerfeld's Elliptical Orbits for Hydrogen[12]

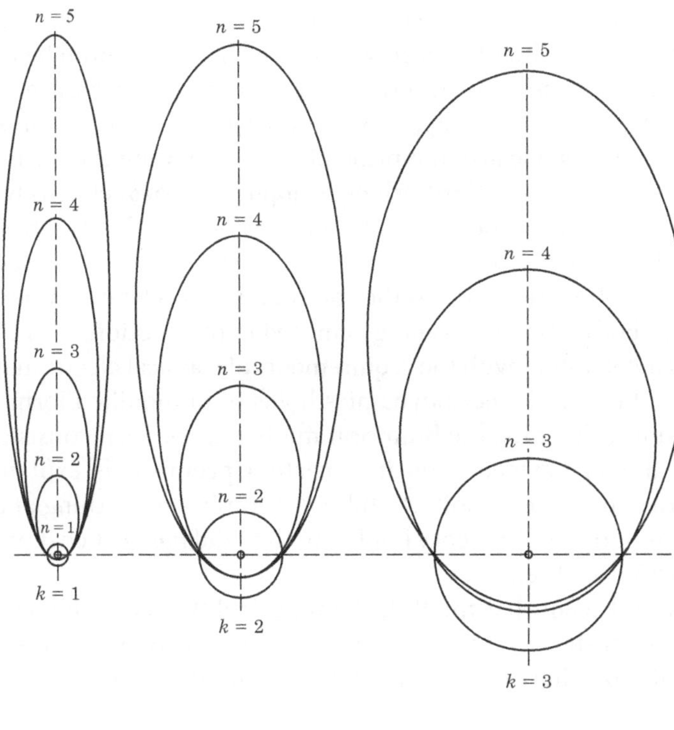

Each Bohr energy level, given by a value of n, actually consists of a number of levels lying close to one another. These microlevels differ in the length of the minor axis of the ellipses (i.e., they differ in the value of the azimuthal quantum number k). For a given n, there are n such sublevels. The number of sublevels increases with increasing n. For $n = 2$, two orbits are permitted, the Bohr circular orbit and an elliptical orbit whose major axis is equal to the diameter of the circle. For $n = 3$, three orbits are permitted, one circular orbit and two elliptical orbits.

Sommerfeld noted that, since the orbital electron's mass is velocity dependent according to the theory of special relativity, its energy in an elliptical orbit differs slightly from its energy in a circular orbit. He observed, in addition, that there should be a slow precession of an elliptic orbit around the nucleus. On the Bohr-Sommerfeld theory of the hydrogen atom, there is a collection of elliptical and circular orbits of different energy between which transitions can be made. Each transition is expressed by a distinct line in the hydrogen spectrum. Wavelengths calculated from these transitions were in good agreement with most instances of observed spectral line splitting (viz., those splitting effects that do not depend on the different energies of opposed electron spin states).

The Bohr-Sommerfeld theory of the hydrogen atom is a causal theory. There are regularities between states of the hydrogen atom before and after energy transitions, and these regularities are reproduced by the theory. Unfortunately, it appears that no explanation is accomplished thereby. This is because the principles on which the theory is based are inconsistent. The third principle applies classical electrodynamic theory to calculate the momentum and energy of the electron in its orbit. The fourth principle stipulates that, even though the electron's circular motion in an orbit is the continuously accelerated motion of a charged body, there is no energy emitted in this motion. This stipulation is inconsistent with the requirements of classical electromagnetic theory. If classical electrodynamics holds in unmodified form in the quantum realm, then the Bohr-Sommerfeld theory is inconsistent.

No theory derived from inconsistent premises is explanatory. A theory that asserts both A and *not* A implies everything and thus explains nothing. If a theory implies both Q and *not* Q, then it provides support for neither.

It would seem that the Bohr-Sommerfeld theory of the hydrogen atom establishes causal relatedness—as regularity between states—without providing explanations of the transitions between sequentially related states. If this indeed is the case, then this successful theory introduces a rift between causality and explanation.

However, a rebuttal may be made to the charge of internal inconsistency. The Bohr-Sommerfeld theory is inconsistent only if it is maintained that the classical electromagnetic theory of moving charged bodies is universally applicable. Bohr himself suggested a way to avoid the assumption of universality. He noted that there is asymptotic agreement between calculations of electron behavior in the quantum theory and calculations from classical electrodynamic theory in the limit as orbital radius increases. Calculations from the equations of motion for discrete electron orbits approach asymptotically calculations for the motion of a free electron.

Bohr elevated this "correspondence principle" to the status of a methodological directive for theories of the subatomic realm. An acceptable quantum theory must display an asymptotic agreement of calculations with the calculations of classical electrodynamic theory.

Since the classical theory of moving charged bodies holds only in the limit at which an electron is no longer bound to a nucleus, there is no inconsistency introduced by maintaining that an orbiting electron does not radiate energy. The Bohr-Sommerfeld theory does not fail to explain because of internal inconsistency. Moreover, the theory does explain why spectral lines appear as they do. The "causality implies explanation" thesis is not threatened by the Bohr-Sommerfeld theory of the hydrogen atom. The causal relations at stake are regularities of transition between electron states, and to establish these regularities is to enable explanatory arguments to be given of atomic spectra. Of course, the Bohr theory is silent about the location of an electron at a particular time either in an orbit or in a jump between orbits. However, this shortcoming, if it is such, does not count against the "causality implies explanation" thesis.

Contributions from Heisenberg, Schrödinger, Born, and De Broglie

The correspondence principle provided a link between discontinuous transitions within the atom to continuous motions adequately described by classical physics. However, Bohr's theory of the atom retained both quantum and classical aspects. Bohr expressed dissatisfaction with his own theory. Henry Folse pointed out that

> Bohr himself cautioned against taking his model as a literal picture of the atomic system not because he suspected the quantum postulate

> was not a true fact of nature or because he wanted to return to some classical representation, but rather because he realized that a single consistent formalism had not been achieved.[13]

Werner Heisenberg developed a candidate for such a formalism in 1925–26. Heisenberg abandoned the concepts of electronic orbit and transition between orbits and replaced them by matrices specifying transition probabilities between stationary states of the atom. Calculations from Heisenberg's matrices were in good agreement with the increasing body of empirical data on atomic spectra. Bohr initially was uneasy about the new matrix mechanics because it appeared to be a purely mathematical device to save appearances.

A few months after Heisenberg applied matrices to describe atomic transitions, Erwin Schrödinger introduced a state function for quantum mechanical systems. Schrödinger's "wave-equation" for one spatial coordinate is

$$\Psi = Ae^{2\pi ih(P\,x - E\,t)}$$

where A is a constant, e is the base of the system of natural logarithms, $i = \sqrt{-1}$, h is Planck's constant, P is a momentum operator, and E is an energy operator.

Given the value of the Ψ-function at initial values of space and time (x_0 and t_0) and the rate of change of Ψ with distance and time, further values of the state of the system may be calculated. The interpretation of an experimental result in the quantum domain is a three-stage process that proceeds from

1. the translation of an initial experimental arrangement into a Ψ-function; to
2. a mathematical analysis of the variation of this function over space and time; to
3. the translation of the results of these calculations into a prediction of the results of a new measurement of the system.

Heisenberg emphasized that quantum mechanical calculations require translations from a language in which observations are recorded into a theoretical language and back again.[14] Max Born suggested a semantical rule for accomplishing these translations: if Ψ represents the state of a subatomic particle, then $|\Psi|^2\, dV$ is the probability that the particle is in the volume element dV.[15] $|\Psi|^2$ is the square of the product of Ψ and its complex conjugate Ψ^*. Complex conjugate Ψ^*, in turn, is formed by substituting ~i for i in function Ψ.

Since $\sqrt{-1^2} = -1$, Born's rule can be used to link Schrödinger's equation, with its imaginary component, to empirically significant quantities such as electron charge densities, orbital transition frequencies and scattering distributions.

In the case of an orbital electron, the Born interpretation correlates $|\Psi|^2$ with electron density. This density is conceived to be a probability distribution of charge within the atom rather than a superposition of discrete orbits. Electron charge density cannot be correlated directly with observed data. However, the distributions predicted from applications of the Schrödinger equation have been confirmed in numerous cases of chemical bonding. The chains of inference that lead from electron charge densities to observable properties of molecules are complex. Nevertheless these chains of inference do end with observables.

In the case of transitions between energy states in atoms, solutions of the Schrödinger equation yield formulas of the form $\Psi = c_1 \psi_1 + c_2 \psi_2 + c_3 \psi_3 + \ldots$, where Ψ is the wave function after transition and c_1, c_2, c_3, ... are amplitudes of the eigenfunctions of the energy states to which the transition may lead. The squares of the respective amplitudes— c_1^2, c_2^2, c_3^2 —specify the probability that the system is in the 1, 2, 3, ... state after the transition. These probabilities are not themselves observables. However, the values of $c_1^2, c_2^2, c_3^2, \ldots$ are correlated with an observable statistical distribution of a great number of systems all undergoing transitions from the same initial quantum state.

In the case of scattering experiments, the Born interpretation correlates $|\Psi|^2$ with the statistical distribution of particles after scattering. It does not correlate $|\Psi|^2$ with the path of any particular particle. The final position of an individual scattered particle is an observable, but no operational procedures are specified to link this position with $|\Psi|^2$. Nevertheless, the Schrödinger equation is applicable to scattering processes that involve a single particle, provided that the single-particle scattering is repeated for a large number of individual incident particles.

As these examples show, there are regularities that link successive states of quantum-mechanical systems. Quantum mechanics and classical mechanics both feature causal laws. The distinction between them is the way in which the state of a system is defined. From the work of Schrödinger and Born, it became clear that a quantum mechanical system can be defined only with reference to probabilities. The probability component is irreducible. This is an important difference from classical mechanics. Classical systems also may be subject to laws of statistical form, but it always is possible to define the states of individual systems that make up a classical ensemble without recourse to considerations of probability. For instance, an ensemble of gas molecules is subject to a Maxwell-Boltzmann statistical distribution of velocities, but instantaneous mass, position, and velocity values for each individual molecule also may be specified, at least in principle.

Since state-function Ψ contains an imaginary component, it is not clear how to interpret the system whose state is specified. Clearly one cannot visualize Ψ as specifying the state of a classical particle or a classical wave.

The problem of interpretation was exacerbated by the suggestion by Louis de Broglie in 1923 that just as light waves are associated with particles (photons), so also particles are associated with waves whose energy is proportional to frequency and whose momentum is inversely proportional to wavelength ($E = hv$; $p = h/\lambda$). He maintained that a beam of electrons should exhibit wave-like behavior in diffraction situations. This prediction was confirmed in experiments conducted in 1926 and 1927. Particularly influential were those of Davisson and Germer (1927), who passed electrons through a crystal of nickel and produced a wave-like diffraction pattern.[16] De Broglie's hypothesis of wave-particle dualism highlighted an important symmetry between these two types of interpretation. He declared that

> if we wish to . . . connect the energy and the momentum of a particle with the frequency and the wavelength which we associate with it, we are led to formulae in which Planck's constant *h* figures in an essential way; this shows that the duality of waves and corpuscles, the necessity of employing two pictures—apparently contradictory—to describe the same phenomenon, is closely associated with the quantum of action.[17]

The two pictures are "apparently contradictory" because the concepts energy and momentum refer to masses localized in small regions of space and time, whereas the concepts frequency and wavelength refer to a wave that is infinitely extended in space and time. It is for this reason that Born declared that

> the imagination can scarcely conceive two ideas which appear less capable of being united than these two, which the quantum theory proposes to bring into such close connection.[18]

This symmetry also is present in the uncertainty relations formulated by Heisenberg in 1927. The uncertainty relations take the form of inequalities, such as $\Delta p \, \Delta q \geq h \,/\, 4\pi$ and $\Delta E \, \Delta t \geq h \,/\, 4\pi$. The first inequality states that the product of the error in measuring the momentum of a particle and the error in measuring its position is equal to, or greater than, $h \,/\, 4\pi$, where h is Planck's constant.

More accurately, it is the *deviations* in a series of measurements of momentum and position that are subject to the limitations specified by the uncertainty relations. The deviation of a set of measurements is the square root of the dispersion of the measurements, and the dispersion, in turn, is the average of the squares of the differences between the individual measurements and the average of the measurements. Deviation is a measure of accuracy of a series of measurements.

The second inequality states that the product of the error, in measuring the energy emitted or absorbed by a body and the error in measuring the time of emission or absorption, is equal to, or greater than, $h \,/\, 4\pi$. The uncertainty relations place limits on the accuracy of simultaneous measurements of position and momentum, or energy and time. The limits are theoretical and not just practical. Indeed, Max Born produced a proof of the uncertainty relations from the formalism of quantum mechanics itself.[19]

Notes

1. C. S. Peirce, "The Doctrine of Necessity," 1892, in *Values in a Universe of Chance: Selected Writing of Charles S, Peirce,* ed. by P. Wiener (New York: Doubleday, 1958), 175.
2. Ernest Rutherford, *Radioactive Transformations* (New Haven: Yale University Press, 1906), 41.
3. Ibid., 14–15.
4. Rutherford, *Radiations from Radioactive Substances* (Cambridge: Cambridge University Press, 1930), 192.
5. Ibid., 192.
6. Ibid., 208.
7. Max Jammer, *The Conceptual Development of Quantum Mechanics* (New York: McGraw-Hill, 1966), 23.
8. Henry Folse, *The Philosophy of Niels Bohr* (Amsterdam: North-Holland, 1985), 61.
9. Gerald Holton, *Introduction to Concepts and Theories in Physical Science,* revised by Stephen G. Brush, 2nd ed. (Princeton: Princeton University Press, 1985), 484.
10. John Heilbron, "Bohr's First Theories of the Atom," in *Niels Bohr: A Centenary Volume,* ed. by A. P. French and P. J. Kennedy (Harvard: Harvard University Press, 1985), 46–47.

11. Arnold Sommerfeld, *Annalen der Physik 51* (1916), 1–94.
12. Gerhard Hertzberg, *Atomic Spectra and Atomic Structure* (New York: Dover, 1944), 18.
13. Folse, *The Philosophy of Niels Bohr*, 66.
14. Werner Heisenberg, *Physics and Philosophy* (New York: Harper, 1958), 46.
15. Max Born, *Atomic Physics* (New York: Hafner, 1946), 92–100.
16. C. J. Davisson and L. H. Germer, "Diffraction of Electrons by a Crystal of Nickel," *Phys. Rev. 30* (1927), 705.
17. Louis de Broglie, *Physics and Microphysics* (New York: Pantheon Press, 1955), 121–22.
18. Born, *Atomic Physics*, 84.
19. Ibid., 381–83.

4

The Copenhagen Interpretation

The Copenhagen interpretation is an interpretation of the quantum domain that emerged from the pioneering work of Heisenberg, Schrödinger, Born, and De Broglie. The integration of these achievements into a systematic view of the quantum realm was primarily the creation of Niels Bohr. For several decades, the Copenhagen interpretation was the "received opinion" about quantum phenomena. It is based on a set of principles:

1. There are three "language levels" for quantum-mechanical interpretation—observational, theoretical, and pictorial.
2. The observational level contains descriptions of experimental arrangements and results.
3. The theoretical level contains statements about the values of state variables for quantum-mechanical systems and the spatiotemporal variations of these variables.
4. A typical quantum-mechanical investigation begins with observational-level descriptions, proceeds to a translation into statements of the theoretical level, continues with a translation of the results of theoretical-level calculations into a prediction of the result of a measurement to be made of the system, and culminates with observational-level descriptions of the results of measurement.
5. Calculations of the theoretical level are applicable only to quantum systems as observed. They provide no account of quantum systems between observations made of them.
6. The pictorial level provides mutually exclusive, but complementary, pictures of the behavior of quantum-mechanical systems between observations.
7. Theoretical-level calculations approach asymptotically the corresponding calculations from classical physics in the limit as the quantum of action ($h = 6.62 \times 10^{-27}$erg.-sec.) becomes negligible.

Observational Level

On the Copenhagen interpretation the observational level is a portion of the language of ordinary experience, augmented by the classical theories of mechanics and electromagnetism. To be included within the observational level, statements must satisfy Bridgman's operational criterion.[1] They must make claims about concepts linked to instrumental procedures that determine their values.

Heisenberg's criticism of Bohr's early theory of the hydrogen atom was based on application of this operational criterion. Bohr's theory postulates that (1) Newton's axioms of motion and Coulomb's law of electrostatic attraction are applicable in the realm of subatomic phenomena; (2) the orbital angular momentum of an electron is quantized in units of $h / 2\pi$; and (3) the energy emitted by an atom in the course of electronic transitions is quantized such that $E = nh\nu$. On the basis of these postulates, Bohr was able to calculate a series of permissible orbits for the electron of the hydrogen atom. Moreover, he was able to render a mathematical account of the observed emission spectra. Heisenberg was not satisfied with this formulation, since Bohr's theory refers to the orbit of an electron, and to the velocity of an electron in this orbit, without regard to the consideration that any attempt to determine the position of the electron in the atom would result in excitation or ionization of the atom. Bridgman pointed out that it was for this reason that

> Heisenberg demanded that only those quantities shall enter the equations which are intrinsically measurable.[2]

The relation of values of a concept to instrumental procedures may be direct or indirect. Values of the concepts "local length" and "local time" are correlated directly with the results of instrumental operations. Other concepts are correlated only indirectly with the results of instrumental operations. An extreme example is the concept "entropy change for an irreversible process," for example, the expansion of a gas into an evacuated container (see above, p. XX). To assign a value to this concept, it is necessary to apply what Bridgman called "paper and pencil operations."[3] To determine the entropy change for an irreversible process, one must devise a method for carrying out the change between the same initial and final states reversibly, and then measure the heat absorbed in the reversible process. This procedure may be indirect, but values of "entropy change for an irreversible process" are determined ultimately by the results of instrumental operations.

Bohr and Heisenberg accepted Bridgman's operational criterion that restricts the content of the observational level to statements about observables. They insisted, moreover, that for a concept to qualify as an observable, it must be susceptible to the measuring procedures of classical physics. Bohr declared that

> notwithstanding refinements in terminology due to accumulation of experimental evidence and developments of theoretical conceptions, all account of physical experience, is, of course, ultimately based on common language, adapted to orientation in our surroundings and to tracing relationships between cause and effect.... The decisive point is to recognize that the description of the experimental arrangement and the recording of observations must be given in plain language, suitably augmented by the usual physical terminology. This is a simple logical demand, since by the word "experiment" we can only mean a procedure regarding which we are able to communicate to others what we have done and what we have learnt.[4]

Heisenberg supported Bohr's analysis. He stated that

> The Copenhagen interpretation of quantum theory starts from a paradox. Any experiment in physics, whether it refers to the phenomena of daily life or to atomic events, is to be described in terms of classical physics. The concepts of classical physics form the language by which we describe the arrangement of our experiments and state the results. We cannot and should not replace these concepts by any others. Still the application of these concepts is limited by the relations of uncertainty. We must keep in mind the limited range of applicability of the classical concepts while using them, but we cannot and should not try to improve them.[5]

It is the Copenhagen position that unambiguous communication about experiments is possible only in terms of the concepts of classical physics. Heisenberg insisted that the Newtonian concepts of space, time, and causality are a priori in the sense that they are universal and necessary conditions for the very possibility of science. They are not, however, of unrestricted applicability. He noted that

> what Kant had not foreseen was that these a priori concepts can be the conditions for science and at the same time can have only a limited range of applicability.[6]

The concepts space, time, and causality must be used in the description of experimental results, but they cannot be applied in the same way to describe what happens between an initial experimental situation and a subsequent measurement made on the system under investigation.

Measuring instruments that are set up for the purpose of observing atomic events are sufficiently heavy to allow an account of their positions and momentums in the terms of classical physics. If the purpose of an experiment is to measure spatial coordinates or time, then rigid, massive measuring rods and clocks are required. If the purpose of an experiment is to measure momentum or energy, the measuring instrument must contain a freely movable part to which the conservation laws may be applied. In either case, the experimental arrangement and the results of observation may be described in the language of classical physics.

Bohr, Heisenberg, and Born assumed that there is a simple, unambiguous correlation between the language of daily life and the concepts of classical physics. Born maintained that

> we are compelled to use the words of common language when we wish to describe a phenomenon, not by logical or mathematical analysis, but by a picture appealing to the imagination. Common language has grown by everyday experience and can never surpass these limits. Classical physics has restricted itself to the use of concepts of this kind.[7]

The concepts of classical physics, then, are merely refinements of the concepts of everyday language. This is one of the most vulnerable claims made by supporters of the Copenhagen interpretation. Studies of the history of the development of such concepts as force and space reveal that the use of these concepts in classical physics differs greatly from their use in everyday life.[8] The refinement that these concepts have undergone in classical physics is quite sophisticated.

Bohr, Heisenberg, and Born may be justly criticized for failing to examine the complex relationship between the concepts of everyday life and the concepts of classical physics. The Copenhagen interpretation can be restated, however, without recourse to the thesis that the concepts of classical physics are refinements of everyday language. What Bohr and Heisenberg were most anxious to demonstrate is that the description of experimental results can be given, and must be given, in the terms of classical physics. The further assertion that the concepts of classical physics are merely refinements of the concepts of everyday life is not necessary to this position.

Theoretical Level

Heisenberg's uncertainty relations form the cornerstone of the theoretical level of the language of quantum mechanics. They place restrictions on the content of the theoretical level, just as the operational criterion places limits on the content of the observational level. No statement that assigns absolute simultaneous values to conjugate quantum mechanical variables, such as position and momentum, is permitted within the theoretical level.

It might seem that this prohibition is overly restrictive. The velocity of an α-particle at the second of two slits, for example, can be defined in terms of the distance between the two slits through which it has passed, and the time taken to traverse this distance. There is no limit to the accuracy with which this determination is made. The same is true of the determination of the position of the second slit. Thus it is possible to calculate values of $\Delta p \Delta q$, which are less than $h / 4\pi$. However, these calculations are of no value for predicting future states of the system. The calculated velocity at the second slit is of no value for predicting the path of the particle upon leaving the slit, because the particle, qua wave, undergoes diffraction at the slit. Given that the purpose of a theory of quantum mechanics is to predict the result of a measurement on the basis of prior measurements made on a system, interpretations that violate the uncertainty relations fail to realize this purpose.

Heisenberg maintained that an interpretation of an experiment in quantum mechanics requires three distinct steps:

1. the translation of the initial experimental situation into a probability function;
2. the following up of this function in the course of time; and
3. the statement of a new measurement to be made of the system, the result of which can then be calculated from the probability function.[9]

Step 1 requires that suitable operators be specified for use in the Schrödinger equation. Schrödinger developed a procedure for assigning operators to classical observables. For one dimension, he replaced observable x by position operator x and observable p_x by momentum operator $h / 2\pi i(\partial / \partial x)$.

Step 2 consists of mathematical operations of either the wave-mechanical calculus of Schrödinger or the matrix algebra of Heisenberg. These two calculi are mathematically equivalent. In step 2, there is no description of what happens to the system under investigation. As Heisenberg expressed this situation,

> it is only in the third step that we change over again from the "possible" to the "actual."[10]

The Born interpretation enables the transition to be made from step 2 to step 3 by correlating the theoretical expression $|\Psi|^2$ with observables, such as scattering distributions and electron transition frequencies. Implementation of the 3-stage procedure outlined by Heisenberg has achieved considerable success in the description and prediction of phenomena in the quantum realm.

In step 1 and step 3 the uncertainty relations place restrictions on the specification of values of conjugate variables, such as position and momentum, or energy and time. According to the Copenhagen interpretation, the uncertainty relations are inextricably bound up with the act of observation. In order to effect a transition from the possible, which is represented by step 2, to the actual, which is represented by step 3, some measurement must be made of the system. This is the only way in which the mathematical calculus can be correlated with events that take place in space and time.

P. W. Bridgman maintained that

> the *modus operandi* by which the uncertainty gets into the situation is through the act of observation . . . the essential fact is that the act of observation interferes with the motion by an unpredictable and uncalculable amount.[11]

Transition from the possible to the actual thus necessarily involves an unpredictable perturbation of the observed system.

The determination of the position of a small particle, for example, an electron, necessitates the use of electromagnetic radiation of very short wavelength. The maximum resolution attainable by means of a microscope is proportional to $\lambda / 2$, where λ is the wavelength of the incident radiation. In order to determine the position of the electron to an accuracy of a few percent, it is necessary to use radiation of wavelength 10^{-10}. Radiation of this wavelength falls within the γ-ray region, and the use of such radiation for the purpose of measuring the position of the electron would necessarily involve the transfer of a large amount of energy to the electron. Because of the large Compton recoil, the momentum possessed by the electron before measurement will be greatly altered by the act of observation. A mathematical analysis of the process yields the formula $\Delta p \Delta q \geq h / 4\pi$.[12] This equation is interpreted by supporters of the Copenhagen interpretation as placing an upper limit on the accuracy of the simultaneous specification of the position and momentum of a particle. The reason that position and momentum cannot be simultaneously determined with, in principle, unlimited accuracy is that the act of measurement necessarily disturbs the system under investigation.

A second experimental example is the determination of the energy of excitation of an atom. In order to determine the exact energy transmitted to an atom, it is necessary to employ monochromatic radiation. But any attempt to ascertain the time at which the absorption of a quantum of energy takes place necessitates a marking of the infinite incident wave train, thereby altering the monochromatic character of the wave. A mathematical analysis of the situation yields the formula $\Delta E \Delta t \geq h / 4\pi$.[13]

A consequence of the Copenhagen interpretation of the uncertainty relations is that every observation of a system that improves the experimenter's knowledge of the magnitude of a certain variable also results in a corresponding decrease in the accuracy with which the magnitude of its conjugate variable is known. The uncontrollable, and not fully predictable, perturbation of a system caused by the act of observation alters the values of previously determined variables.

The Copenhagen interpretation places an additional restriction on theoretical-level interpretations (over and above the restrictions placed by the uncertainty relations). This additional restriction is the correspondence principle that Bohr had invoked in developing his 1913 theory of the hydrogen atom. In its initial version, the correspondence principle stipulates that there be asymptotic agreement in calculations with the equations of classical electrodynamic theory as the radius of the hydrogen electron's orbit increases.

Encouraged by the success achieved by the theory of the hydrogen atom, Bohr promoted the correspondence principle to the status of a criterion of acceptability for any theory of the quantum realm. Any acceptable theory of the quantum realm must establish asymptotic agreement with classical theory for those applications in which the classical calculations are empirically adequate.

The calculations of both quantum theory and special relativity theory exhibit asymptotic agreement with the calculations of classical theory. The Schrödinger equation for a bound electron reduces to the classical equation for the orbit of an electron revolving around a positively charged nucleus of the same charge value in the limiting case of an infinitely large electron radius. Similarly, the equations of special relativity theory yield the same results as the Newtonian equations of motion in the limiting case in which velocities are negligible compared to the velocity of light. These instances of asymptotic agreement presumably render quantum theory and special relativity theory more acceptable than if there were no such agreement.

Several philosophers of science have supported Bohr's recommendation that the correspondence principle be taken to be a criterion of acceptability for theory replacement. Joseph Agassi insisted that it is an "acknowledged methodological demand" that a newly proposed theory "should yield the theory it comes to replace as a consequence or as a first approximation and also as a special case."[14]

Ernest Hutten maintained that

> since the [correspondence] principle states that the new theory contain the old theory as a special case, it prescribes the conditions the new calculus must satisfy; and the possibilities for constructing new theories are curtailed in this manner. Thus, the correspondence principle safeguards the relation to experiment.[15]

Agassi and Hutten interpret the correspondence principle to require the incorporation of a predecessor theory into its successor. The incorporation view of scientific progress has been an influential position within the history of the philosophy of science. Its author was William Whewell (1794–1871), who likened the historical development of science to a confluence of tributaries to form rivers. This confluence is an incorporation of past results into present theories in such a manner that unification is achieved. Whewell listed three conditions that are individually necessary and jointly sufficient for progressive theory replacement: (1) the successor theory is consistent; (2) the successor theory is more inclusive than, and incorporates, the predecessor theory; and (3) the gain in inclusiveness is accompanied by a gain in simplicity. The mere conjoining of theories is not sufficient.[16] What is required is an integration that unifies an expanded realm of experience. Of course, to require integration and unification is not much more definite than to insist that simplicity be achieved. Whewell did not formulate a satisfactory definition of simplicity.

According to Whewell, the history of astronomy is a successful realization of these conditions. Results achieved by Copernicus, Galileo and Kepler were incorporated into the more inclusive theory of Newton. This incorporation with unification was based on a superposition of certain ideas—the three axioms of motion and the law of gravitational attraction—upon "facts" established by Copernicus, Galileo, and Kepler.

Newton's synthesis provided a rationale for Kepler's third law, $(T_1/T_2)^2 = (D_1/D_2)^3$, where T is the period and D is the distance of a planet from the sun. Kepler presented the law as a mathematical device to save appearances. He provided no reason why planetary periods and distances should be related in this way. Newton demonstrated that this relation follows from the idea of inertial motion and the idea of $1/R^2$ forces directed toward the sun, considered to be a stationary point-center of force. Newton's theory not only accounted for the approximate fit between Kepler's third law and the motions of the planets, but also explained—via perturbation theory—why, and to what extent, the agreement is only approximate.

In the twentieth century, Ernest Nagel and Imre Lakatos presented influential versions of the incorporation view. Nagel noted that there are instances in the history of science in which one theory is replaced by a more inclusive theory that subsumes it. The paradigm case is the reduction of classical thermodynamics to kinetic molecular theory.

Nagel formulated a set of formal and empirical criteria for the replacement of one theory by a second more inclusive theory.[17] There are two formal criteria of successful reduction: (1) each term of the reduced theory (e.g., temperature) is linked to terms of the reducing theory (e.g., $\frac{1}{2}\, nmu^2 = kT$), and (2) the laws of the reduced theory are derivable from the interpreted axioms of the reducing theory. There also are two empirical criteria of successful reduction: (1) the principles of the reducing theory are supported by empirical evidence over and above that evidence that supports the reduced theory, and (2) the development of the reducing theory results in a further growth of scientific knowledge.

Paul Feyerabend argued that the historical examples put forward by supporters of progress by reduction fail to satisfy the above criteria. According to Feyerabend, it is partial overlap and not incorporation that is the normal pattern within the history of science.[18]

Imre Lakatos restated Whewell's "progress through incorporation" view within the context of unfolding scientific research programs. A scientific research program consists of a central core of axioms and principles, augmented by a collection of auxiliary hypotheses that facilitate its application to experience. Examples of such programs are (1) the Newtonian program for planetary astronomy; (2) the Cartesian program to account for all motion by reference to impact or pressure; (3) the LaPlacian program to explain electrical, magnetic, and chemical phenomena by postulating central-force interactions between particles of ponderable matter and particles of imponderable matter; and (4) the Proutian program to show that the atomic weights of the chemical elements are exact multiples of the atomic weight of hydrogen.

Lakatos specified a criterion of theory replacement within the development of a scientific research program. The replacement of theory T_n by theory T_{n+1} is progressive provided that

1. T_{n+1} has greater empirical content than T_n;
2. T_{n+1} accounts for the prior successful applications of T_n; and
3. Some of the excess content of T_{n+1} is corroborated.[19]

Pictorial Level

N. R. Hanson maintained that one important feature of the Copenhagen interpretation is Bohr's insistence that

> all empirically testable statements within quantum theory are ultimately about "phenomena" (i.e., the-specimen-particle-plus-the-detecting-apparatus-without-the-mutual-interaction-of-which-no-elementary-particle-ever-can-be-apprehended).[20]

In the case of electron diffraction at a slit, empirically testable statements are restricted to the electron-slit interaction and the electron-photographic-plate interaction. No empirically significant statements can be made about an electron between slit and plate. One may claim that its state is specified at every point by the Schrödinger equation, but since this equation contains an noneliminable imaginary component ($\sqrt{-1}$), the claim lacks empirical significance.

Bohr maintained, nevertheless, that meaningful statements can be made about quantum systems between observations made of them. These "pictures" do not enter into theoretical-level calculations of scattering distributions or electron transition frequencies. Rather, they are superimposed upon the theoretical-level results so as to provide descriptions of the behavior of quantum systems between observations.

Pictorial-level interpretations may take the form of mutually exclusive descriptions of a quantum system in terms of classical particles or classical waves. The two pictures cannot both be applied to a particular experimental result. The theorist may superimpose the particle picture or the wave picture, but not both. It is the Copenhagen position that the two mutually exclusive pictures are complementary. A *complete* explanation of a quantum phenomenon requires superposition of both pictures.

Consider the diffraction of electrons at a single slit. Bohr noted that it is possible to set up apparatuses to measure accurately either the path of a particle or the diffraction pattern (actually, no measurements are made on the particle between the slit and the photographic plate upon which the particle makes impact). If the diaphragm is rigidly connected to a massive frame, the transverse component of the position of a particle passing through the diaphragm can be determined by means of the position of the slit. The transverse component of the momentum of the particle then may be determined with only low accuracy, although the distribution of particles striking the photographic plate can be calculated from the Schrödinger equation upon substitution of a specific distance between diaphragm and plate. At the pictorial-language level, the wave picture is appropriate. The pattern on the photographic plate is explained by reference to the diffraction of a wave at the slit. Bohr pointed out that if the diaphragm instead is connected to the frame by means of an elastic spring,

> it should, in principle, be possible to control the momentum transfer to the diaphragm and, thus, to make more detailed predictions as to the direction of the electron path from the hole to the recording point.[21]

This increase in accuracy with which the path of a particle may be determined is achieved only at the expense of a blurring of the diffraction pattern. At the pictorial-language level, the particle picture is appropriate. The electron is pictured as a particle whose path is determined by the momentum transfer at the slit.

The experimental arrangement in which an interpretation in terms of "wave language" is applied (the diaphragm is rigidly connected to the frame), does not lend itself to an interpretation in terms of "particle language." Conversely, the experimental arrangement in which an interpretation in terms of particle language is applied (the diaphragm is connected to the frame by an elastic spring), does not lend itself to an interpretation in terms of wave language. Bohr placed a restriction on linguistic usage at the pictorial level. The results of a particular experiment may not be interpreted in terms of both a particle picture and a wave picture. Thus, although the wave picture and the particle picture are mutually exclusive descriptions of what happens between measurements made on a system, the pictorial level of language does not contain logically incompatible claims. In the case of the passage of electrons through a single slit, for instance, the Copenhagen interpretation restricts the wave picture to one experimental arrangement and the particle picture to a second experimental arrangement. Given a specific experimental arrangement, it does *not* follow on the Copenhagen interpretation that both a wave passed through the slit and a particle exchanged momentum with the slit.

In addition to restricting the interpretation of a given experiment to one or the other of two mutually exclusive descriptions, Bohr's principle of complementarity specifies that two mutually exclusive descriptions of a quantum phenomenon, in which two different experimental arrangements are employed, are complementary and combine to produce an exhaustive interpretation of the phenomenon. In the case of the diffraction of electrons by a single slit, the two mutually exclusive descriptions are (1) electrons, which have a certain position in passing through a diaphragm, produce a certain pattern of impacts on a photographic plate, and (2) electrons, which have a certain momentum in passing through a diaphragm, transfer a certain momentum to the diaphragm. These two descriptions combine to exhaust the range of possible knowledge about the passage of electrons through a slit in a diaphragm. To fully explain the results of single-slit experiments, it is necessary to superimpose both particle and wave pictures.

Bohr maintained that mutually exclusive, but complementary, pictorial-level descriptions also may be given in terms of the paths followed between observations, and causal interactions in which energy and momentum are exchanged. C. A .Hooker and Henry Folse have emphasized that Bohr's primary emphasis was that spatiotemporal description and causal interaction are mutually exclusive, but complementary, pictorial-level interpretations, and that it was of lesser importance to Bohr that particle pictures and wave-pictures also are mutually exclusive, but complementary pictorial level interpretations.[22]

At the pictorial level of interpretation, causal descriptions are superimposed upon quantum processes. Unfortunately, the "causal attribution implies explanation" thesis cannot be retained at this level. The causal picture based on momentum-energy exchange is applicable only to the extent that spatiotemporal description is inapplicable. People believe that they understand the collision of two billiard balls because they can apply both conservation principles to the impact and trace the path of the balls over time. For an interaction at the quantum level, by contrast, demands for an explanation of the process are thwarted by the prohibition against combining causal analysis and path description for a particular experimental result.

Critics of the Copenhagen interpretation take the failure of the "causal attribution implies explanation" thesis at the pictorial level to be a good reason to eliminate this level of interpretation. Calculations at the theoretical level suffice for the prediction of experimental results. Pictures that "describe" what happens between observations are irrelevant to this task.

Notes

1. P. W. Bridgman, *The Logic of Modern Physics* (New York: Macmillan, 1927), 1–32; *The Nature of Physical Reality* (Princeton: Princeton University Press, 1936), 5–15.
2. Bridgman, *The Nature of Physical Reality,* 65.
3. Bridgman, *Reflections of a Physicist* (New York: Philosophical Library, 1950), 15.
4. Niels Bohr, "Quantum Theory and Philosophy," in *Philosophy in the Mid-Century,* ed. by R. Klibansky (Firenze: La Nuova Italia Editrice. 1958), 308–10.
5. Werner Heisenberg, *Physics and Philosophy* (New York: Harper & Brothers, 1958), 44.
6. Ibid., 90.
7. Max Born, *Atomic Physics,* (New York: Hafner, 1946), 92.
8. See, for instance, Max Jammer, *Concepts of Force* (Cambridge: Harvard University Press, 1957), and *Concepts of Space* (Cambridge: Harvard University Press, 1959).
9. Heisenberg, *Physics and Philosophy,* 46.
10. Ibid., 47.
11. Bridgman, *Reflections of a Physicist* (New York: Philosophical Library, 1955), 222.
12. See, for instance, Max Born, *Atomic Physics,* 94–95.
13. Ibid., 96.
14. Joseph Agassi, "Between Micro and Macro," *Brit. J. Phil. Sci. 14* (1963), 26.
15. Ernest Hutten, *The Language of Modern Physics* (London: George Allen & Unwin, 1956), 168.
16. William Whewell, *Philosophy of the Inductive Sciences, Part II* (London: Cass, 1967), 85.
17. Ernest Nagel, *The Structure of Science* (New York: Harcourt, Brace & World, 1961), 345–61.
18. Paul Feyerabend, "Explanation, Reduction, and Empiricism," in *Minnesota Studies in the* Philosophy *of Science, III* (Minneapolis: University of Minnesota Press, 1962), 46–59; 76–80.
19. Imre Lakatos, "Falsification and the Methodology of Scientific Research Programmes," in *Criticism and the Growth of Knowledge,* ed. by I. Lakatos and A. Musgrave (Cambridge: Cambridge University Press, 1970), 116–34.
20. Norwood Russell Hanson, "Five Cautions for the Copenhagen Interpretation's Critics," *Phil. Sci. 26* (1959), 327.

21. Bohr, *Atomic Physics and Human Knowledge* (New York: Wiley, 1958), 45.
22. C. A. Hooker, "The Nature of Quantum Mechanical Reality," in *Paradigms and Paradoxes*, ed. by R. Colodny (Pittsburgh: University of Pittsburgh Press, 1972), 140, 158;
Henry Folse, *The Philosophy of Niels Bohr* (Amsterdam: North-Holland, 1985), 222–57.

5

The Rift between Causality and Explanation

The Einstein-Podolsky-Rosen Challenge

In 1935, Einstein, Podolsky, and Rosen (EPR) argued that the Copenhagen interpretation of the quantum realm is incomplete. Moreover, they maintained that it is based on an unacceptable interpretation of physical reality.[1]

The EPR paper directs attention to a thought experiment in which two subatomic particles interact and then separate. Before interaction, both system 1 and system 2 are fully described by individual wave functions. After interaction, there is a composite wave function for the combined system (1 + 2).

Suppose the post-interaction position of system 1 is determined experimentally. Given this information, the resources of the theoretical level (e.g., the wave function for the combined system) can be utilized to calculate the post-interaction position of system 2. One also can calculate, from the post-interaction momentum of system 1 and the composite wave function, the post-interaction momentum of system 2. Thus one may assign values of position and momentum to system 2 after interaction without physically interfering with the system. Of course, any direct simultaneous measurement of position and momentum of system 2 after interaction would be subject to the restriction imposed by Heisenberg's uncertainty principle. However, since no such measurement is made in the EPR thought experiment, the accuracy of the values of position and momentum assigned to system 2 after interaction is unlimited.

It is the EPR position that

> if, without in any way disturbing a system, we can predict with certainty (i.e., with a probability equal to unity) the value of a physical quantity, then there exists an element of physical reality corresponding to this physical quantity.[2]

It has been shown that values of the position and momentum of system 2 after interaction can be calculated, with no restriction on accuracy, without interfering with the system. Hence this system possesses a definite position and momentum, the values of which are not subject to the restriction of the uncertainty principle. Since the Copenhagen interpretation does not possess the resources to specify these values, it must be incomplete as a theory of the quantum realm.

Bohr replied that the EPR calculations are paper-and-pencil operations that have no empirical value. On the Copenhagen interpretation, one cannot specify the state of each system separately after interaction. One only can specify the state of the combined system. The only way actually to determine the value of the position (or momentum) of system 2 after interaction is to perform a measurement on that system. And a real measurement, as opposed to a theoretical calculation, is subject to the restriction placed by the uncertainty principle on the simultaneous determination of position and momentum.

Neutral Pion Decay and the "Causality Implies Explanation" Thesis

David Bohm revised the EPR thought experiment by focusing on spin determination as a state variable.[3] Consider the creation of a pair of electrons by the spontaneous decay of a neutral pion. This is a physical process whose existence has been established. The spin of the pion before decay is zero. Consequently, the sum of the spins of the electrons also must be zero, given that spin is conserved in nuclear reactions.

One can measure the spin direction of one of the electrons in the vertical plane at right angles to its motion. The result will be "spin up" or "spin down" depending on the electron's spin direction. Alternatively, one can measure the electron's spin direction in the horizontal plane, in which case the result will be "spin right" or "spin left."

The investigator has a choice. One can measure the spin direction of an electron in either the vertical plane or the horizontal plane, but not both. This is an application of Heisenberg's uncertainty principle. As noted above, the uncertainty principle places limitations on the accuracy with which the simultaneous position and momentum of a particle may be determined. In this application, this is a relative restriction, in the sense that the more accurately position is determined, the less accurately momentum can be determined and vice versa. In the case of spin momentum, the restriction is absolute. If one determines the spin direction of a particle in the vertical direction, one cannot determine the spin direction at that time in the horizontal direction.

Suppose the choice is made to measure the vertical spin of electron #1. This may be accomplished by placing a Stern-Gerlach magnet in its path. If the result of measurement is "spin up," it follows from the principle of conservation of spin that the vertical spin of electron #2 is "spin down." This is the case even before making a measurement on electron #2. David Lindley noted that

> it seems that a spin measurement on one electron has in effect "measured" the spin of the second, remote electron, without any actual physical intervention. The second electron has been reduced from a wholly indefinite unmeasured state to a definite "down" state, even though nothing has been directly done to it.[4]

The pion-decay sequence is a causal relation on the regularity-between-states view. One can predict successfully the state of a system—the spin status of electron #2—from a regularity governing state descriptions and an initial measurement.

The normal relation between causal attribution and scientific explanation is not present in this case, however. The measured result of spin determination on electron #1 (spin up) may be the cause of the down value for electron #2, but, if this is so, this is a case of instantaneous action at a distance. The theory of special relativity stipulates that no physical influence can be propagated with a velocity greater than the velocity of light. If this theory is correct, then to affirm that both causal interaction and action at a distance takes place is to affirm a self-contradiction, and hence fail as an explanation.

Two-Slit Experiments and the "Causality Implies Explanation" Thesis

Analysis of the two-slit experiments also reveals a failure of the causality-explanation relationship. If both slits are open, a beam of electrons directed at the diaphragm at the midpoint between the slits displays a typical interference pattern on striking a photographic plate behind the diaphragm. The same pattern is produced if a large number of electrons of the same energy are fired, one by one, at the diaphragm.

Figure 11
Passage of Electrons through Two Open Slits

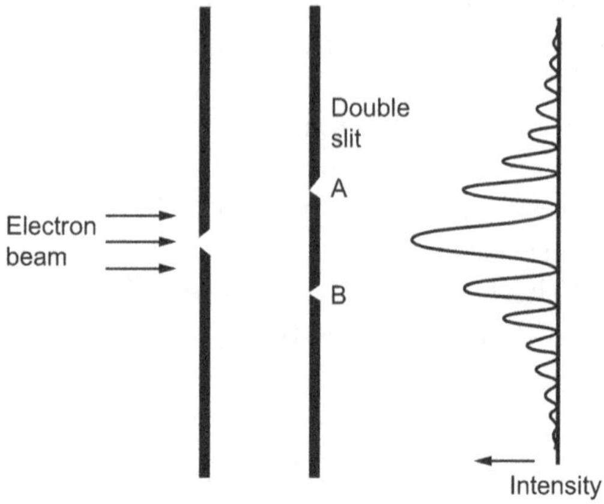

It would seem that in this experiment, electrons manifest wave-like properties. Electron waves that originate at slit *A* interact with electron waves that originate at slit *B* to produce a diffraction pattern. If slit *A* is closed, electrons that pass through slit *B* create the bell-shaped distribution pattern below.

Figure 12
Passage of Electrons through a Single Slit

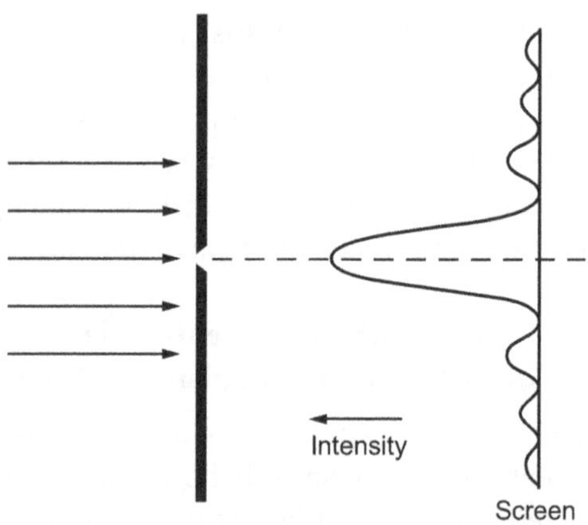

It would seem that, in this modified experiment, electrons manifest particle-like properties as they pass through slit *B*. Consider again an experimental arrangement in which slit *A* and slit *B* both are open, and a detector is present at each slit. This is a different experimental arrangement from the case in which detectors are not present. On the Copenhagen interpretation, the physical system and the recording instruments must be treated as an indivisible whole. In Leon Rosenfeld's picturesque imagery, there is no "view from Sirius" from which a system can be observed without interacting with it.[5] What is observed on the photographic plate when electrons are detected passing through one slit or another is a juxtaposition of individual bell-shaped curves from each slit.

Figure 13
Passage of Detected Electrons through Two Open Slits

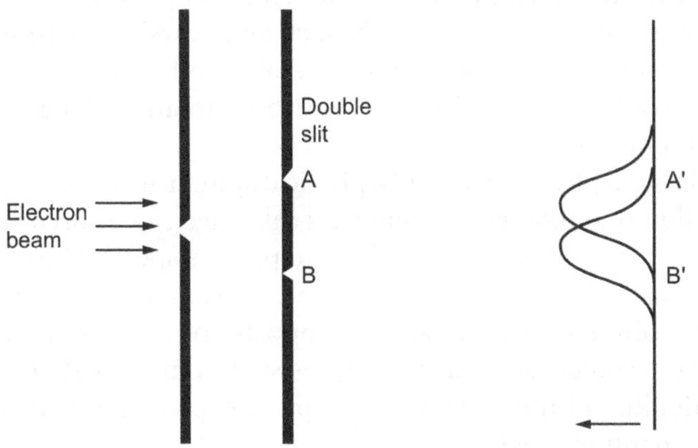

The interference effect produced when no detecting devices are present at the two slits is no longer present. It would seem that detecting the passage of individual electrons through the slits destroys the wave-like behavior of the electrons.

The situation is further complicated if slit *A* and slit *B* both are open, but a measuring device is present only at slit *A*. Surprisingly, the pattern displayed on the screen is a juxtaposition of the bell-shaped distributions from slit *A* and slit *B*, and not the interference pattern produced when no detectors are present.

Electrons that pass through slit *B* respond to the fact that electrons have been recorded passing through slit *A* and also act as particles. John Gribbin noted that "somehow, the electrons going through the second hole 'know' that we are looking at the other hole, and also behave like particles as a result."[6]

If regularity between states of a physical system counts as causal relatedness, then the two-slit experiment, with both slits open and no detectors at the slits, is a causal process. The interference pattern may be calculated from the principles of quantum theory. There is available a deductive-nomological pattern of explanation, with a causal law as major premise and a conclusion that describes the interference pattern.

However, this relation between causality and explanation precludes the ascription of continuous spatiotemporal paths to the particles. Each individual electron passes through the diaphragm and strikes the photographic plate such that it contributes to the overall interference pattern. One cannot, without destroying the interference pattern, determine whether an electron passes through slit *A* or slit *B*. It would seem that each electron knows where to strike the plate in order to contribute to the overall interference pattern. Does this "it would seem that" claim contribute to an explanation of the experimental results?

In classical physics, states of a physical system are causally related if they follow one another according to a rule (given appropriate isolation of the system). Causal laws in classical physics specify the successive spatial positions of a system over time. Explanations of processes in classical physics typically are a composite of separate accounts of energy or momentum transfer and the spatiotemporal description of the unfolding of the system. This type of explanation is unavailable within quantum mechanics.

Until the development of quantum theory, it was understood that causal explanation is set within a context of spatiotemporal description. Examples include applications of the laws of conservation of momentum and conservation of energy to the motion of balls on pool tables, and applications of the laws of Hooke and Newton to the simple harmonic motion of objects released at the ends of stretched springs.

Quantum theory affects a divorce between causal accounts and spatiotemporal descriptions of physical systems. If one insists that a concomitant spatiotemporal description is a requirement for causal explanation, then it is false that successful causal attribution implies explanation.

Notes

1. Albert Einstein, Boris Podolsky and Nathan Rosen, "Can Quantum Mechanical Description of Reality Be Considered Complete?" *Phys. Rev.* 47 (1935), 777.
2. Ibid.
3. See David Lindley, *Where Does the Weirdness Go?* (New York: Basic Books, 1996), 96–100.
4. Ibid., 97.
5. Leon Rosenfeld, "Misunderstandings about the Foundations of Quantum Theory," in *Observation and Interpretation*, ed. by S. Korner (New York: Academic Press, 1957), 42.
6. John Gribbin, *Schrödinger's Kittens and the Search for Reality* (Boston: Little, Brown and Co., 1995), 13.

6

The Rift between Causality and Explanation: Response Options

The neutral pion experiment and the two-slit experiments pose difficulties for the thesis that causality implies explanation. There are a number of options available.

Option One: Affirm that the "Causality Implies Explanation" Thesis Holds for the Results of Measurements

One option is to hold that the "causality implies explanation" thesis does hold for relations between quantum phenomena, provided that quantum phenomena are identified with the results of measurements. It is incontrovertible that there are regularities between quantum phenomena. These regularities result from interactions between measuring instruments and quantum systems.

The neutral pion experiment is an example. In this experiment there is an observed regularity between the result of measuring the spin direction of electron #1 after emission and the result of measuring the spin direction of electron #2.

The single-slit experiment is another example. In this experiment, there is an observed regularity between electron beams of specified measured energy and direction and impact patterns on photographic plates.

If causal relatedness is interpreted to be a Humean constant sequential conjunction between members of two classes of events, then the neutral pion experiment and the single-slit experiment are genuinely causal. To defend this position is to shift causal relatedness from regularity between states of physical systems to regularity between events (qua measurements made of physical systems by measuring instruments under specified conditions.).

What are to be explained in option 1 are regularities between events. There are available sound deductive-nomological arguments that explain the regularities discussed above. In the case of the neutral pion experiment, the premises include the law of conservation of spin, a description of pion decay, and the result of measuring the spin direction of electron #1. The conclusion of the argument is a description of the result of a measurement of the spin direction of electron #2.

In the case of the single-slit experiment, the premises include the mathematical formulism of quantum theory, a description of the experimental arrangements, and the initial measurement of the electron beam. The conclusion of the argument is a description of the impact pattern on the photographic plate.

The arguments that explain the results of measuring the spin direction of electron #2 and the impact pattern from electron diffraction have nomological force because the conservation of spin direction and the Schrödinger equation presumably are genuine laws. From the standpoint of option 1 there is no challenge to the "causality implies explanation" thesis.

A similar analysis is appropriate on an "inferability theory" of causality. If inferability is taken to be a sufficient condition of causality, then numerous experiments in the quantum domain are causal. Philipp Frank developed an inferability version of causality. His succinct statement of this position is that a relation is causal if it

> allows us to infer from information about one region of space and time some information about another region of space and time.[1]

Frank applied the inferability theory to successive states of (approximately isolated) physical systems and not to Humean regularities between events.[2] However, if inferability is taken to be a sufficient condition of causal relatedness of the members of two classes of events, then the neutral pion experiment and the various electron diffraction experiments are events that are both causally related and subject to deductive-nomological explanations.

Deductive-nomological explanatory arguments aside, unease returns when refocusing from correlations between recorded experimental results to the behavior of quantum systems themselves. Demands for explanation are thwarted because the neutral-pion and two-slit experiments exhibit an instantaneous transmission of influence from one location to another, a transmission that violates a basic postulate of the theory of special relativity. One cannot explain an experimental result by claiming both that an influence is transmitted instantaneously from point A to point B, and that no influence can be transmitted over distance instantaneously. Whatever theory of explanation is adopted, it is clear that arguments based on inconsistent premises fail to explain.

There is an additional difficulty for the "causality implies explanation" thesis, even if one restricts attention to correlations between the results of measurement. Suppose an electron passes through a slit in a diaphragm and strikes a photographic plate. One can predict, from the formalism of quantum theory, the probability that the electron strikes the plate within a specified area. Suppose this probability is 2 percent. One cannot explain the fact that a single electron did strike the plate within this area. Given a probability of 2 percent, what would be better explained is the absence of an impact within the area in question.

Of course, if a large number of electrons pass through the slit, the calculated probability values do correspond to the statistical distribution of impacts on the plate. It is appropriate to claim that the theory does explain the distribution of impacts on the plate. One conclusion to be drawn from investigations of the quantum realm is that expectations about what can be explained must be reassessed. The behavior of an ensemble of particles may be explained even though the behavior of an individual particle is not subject to explanation.

Option Two: Causality Holds at a Sub-quantum Level

A second option is to retain the "causality implies explanation" thesis by locating causal relatedness at a deeper level of interaction. Just as the phenomenal laws of classical thermodynamics have a "genuinely causal" basis in kinetic molecular theory, so also quantum mechanics may have a genuinely causal basis in interactions at a sub-quantum level.

This approach was selected by David Bohm and Jean-Pierre Vigier. Bohm and Vigier conceded that the theory of Bohr, Heisenberg, and Born accounts for numerous types of regularity in the quantum realm. It is a causal theory on the regularity-between-states interpretation of causality. Unfortunately, the regularities developed within standard quantum theory do not facilitate satisfactory explanations of the relevant phenomena. There is a rift between causal relatedness and the explanation of quantum processes. Perhaps this relationship can be reestablished by reference to a deeper level of interaction.

Bohm pointed out that other interpretations of the mathematical calculus are not excluded by the calculus itself. He suggested that

> one might assume that the indeterminacy in the position and momentum of the electron is a consequence of the fact that these variables, which after all are suggested only by the extrapolation of classical physics to the atomic domain, are not a complete description of the electron.[3]

Bohm insisted that Von Neumann's proof that there can be no hidden variables, which would enable a more accurate description of the state of a quantum mechanical system than is possible on the Copenhagen interpretation, assumes that

> at least part of the specification of the state of the system will always be in terms of these observables, while the hidden variables will at most serve to make more precise the specification already given by the observables. Such an assumption evidently severely limits the forms of the theories that may be taken into consideration. For it leaves out the important possibility that as we go to a sub-quantum mechanical level the entire scheme of observables satisfying certain rules that are appropriate to the quantum-mechanical level, will break down, to be replaced by something very different.[4]

Bohm and Vigier hypothesized that there is a sub-quantum level comprising micro-objects that exist independently of all observation. These objects possess both particle-like and wave-like aspects. Bohm noted that

> in previously existing theories only two possibilities were considered, namely, that of the pure wave and that of the pure particle, these two possibilities being regarded as mutually exclusive. On the other hand, it is evidently possible that in any given process both wave and particle could be present *together* in some kind of interconnection.[5]

Given that such micro-objects exist, Vigier declared that

> our problem is then to indicate a model of individual micro-objects, which implies as statistical mechanics for its particle-like aspect precisely the statistics of quantum mechanics.[6]

Bohm and Vigier thus accepted Bohr's correspondence principle as a methodological directive. They insisted that an acceptable "deeper-level" theory must achieve asymptotic agreement of calculations with standard quantum theory in that region for which the standard theory has proved effective.

Bohm suggested that a deeper level of interaction underlies quantum phenomena in much the same way that atomic collisions underlie Brownian motion.[7] This deeper level of interaction specifies the precise details of the motions of individual electrons and photons, subject to the restriction that there is asymptotic agreement of calculations with the results achieved by the standard theory in the quantum realm.

Bohm hypothesized that each individual electron and photon is "connected with" a micro-object.[8] Each micro-object is associated with a wave whose mathematical representation is the Schrödinger Ψ-function. In this interpretation, the Ψ-function is an objectively real wave. The Ψ-field and the micro-object are assumed to exert a reciprocal force on one another at the sub-quantum level. At the quantum level, the influence of the Ψ-field on the micro-object is dominant; the reciprocal influence of the object on the field is negligible.

Bohm and Vigier assumed that the Ψ-field undergoes random fluctuations about an average value, which average value is given by the Schrödinger equation. The micro-objects are presumed to be influenced by the minute fluctuations in the Ψ-field. Given suitable assumptions about the deeper-level fluctuations, the Born probability distribution $|\Psi|^2$ follows. Bohm noted that the Born distribution is a deductive consequence of the principles of his theory of the sub-quantum domain. By contrast, the Born probability distribution functions as an arbitrary supposition on the Copenhagen interpretation. Bohm declared that

> instead of starting from Born's probability distribution as an absolute and final and unexplainable property of matter, we have shown how this property could come out of random motions originating in a sub-quantum mechanical level.[9]

It might seem that Bohm's theory of a sub-quantum domain restores the "causality implies explanation" thesis on the quantum level. Consider the two-slit experiment: Bohr and Heisenberg had shown that the statistical distribution of impacts on the photographic plate conforms to what is predicted from the Born interpretation of the formalism. The predicted distribution is observed regardless of whether a beam of particles is directed on the diaphragm or individual particles pass through the slits one by one. However, there is no satisfactory explanation available for the effect of observing the passage of particles through slit *A*. It is puzzling that particles passing through slit *B* apparently know about the observation at slit *A*, and as a result, modify their contribution to the pattern of impacts.

Bohm and Vigier sought to describe the behavior of individual micro-objects. According to Bohm and Vigier, the particle-like aspect passes through one slit only, and the wave-like aspect passes through both slits. After passing through one of the slits, the particle-like aspect is influenced in its motion by the diffracted wave-like aspect in such a manner that the micro-object is most likely to strike the plate at positions corresponding to maxima in the interference pattern.

The micro-object connected with each electron possesses a random motion, such that its path through slit *A* is irregular. If both slits are open, the cumulative electron impacts display an interference pattern.

Figure 14
Bohm Theory—Both Slits Open[10]

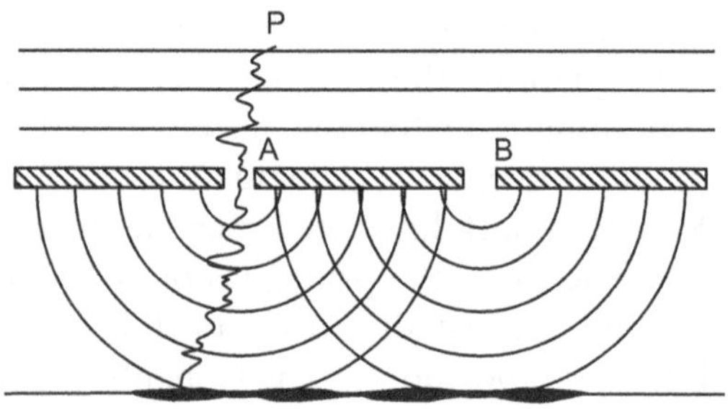

If slit *B* is closed, the quantum force on the micro-object is altered, and the cumulative electron impacts display a bell-shaped distribution.

Figure 15
Bohm Theory—Slit B Closed[11]

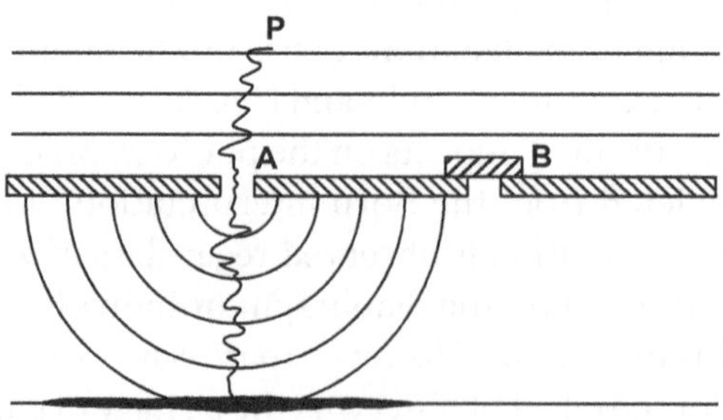

Bohm maintained that, on his theory of sub-quantum interactions,

> we can understand how the wave-particle duality originates. . . . This result was achieved by the simple expedient of considering that the electron is a combination of particle and field, interconnected and undergoing suitable random fluctuations in its motion.[12]

Bohm insisted that, on the Copenhagen interpretation, "no such understanding is possible."[13] The Copenhagen interpretation only provides mutually exclusive wave and particle pictures, each applicable to a distinct experimental arrangement. Bohm held that since the theory of sub-quantum interactions preserves the "causality implies explanation" thesis, and the Copenhagen interpretation does not, the former theory is preferable.

But does it count as an explanation to attribute the results of the two-slit experiment to objects that qua particle pass through one slit but that qua wave pass through both slits? A particle that also is a wave is a prima facie self-contradiction. Particularity implies specific spatial location at a given time; "wave-ness" implies unlimited spatial extendedness at a given time.

Critics have leveled two types of complaint against Bohm's theory. Some critics have pointed out that there is no experimental confirmation for the theory over and above the evidence that already supports orthodox quantum theory. This type of criticism leaves open the possibility of confirmation at some future time. Bohm himself held out the possibility that experiments in the high energy–short time interval domain will provide support for his theory.

A second type of criticism is that Bohm's theory is unconfirmable. D. M. MacKay emphasized that information about the physical world can be obtained only in the course of an interaction between the object under investigation and the measuring instruments.

> Does Professor Bohm think that it will be possible to start with an instrument whose quantum of action of perturbation is h, and end up with information about action in units smaller than h? If the basic unit of interaction is the unit h (which is still an empirical question), then all the information patterns which we build to describe the world will also be quantized in bricks of h as far as they are operationally defined.[14]

According to the Copenhagen interpretation, any "hidden variables" introduced into quantum theory must be linked to observables through the uncertainty principle, which governs the accuracy with which certain combinations of observables may be specified. The uncertainty principle, coupled with the requirement that any interpretation of quantum phenomena be linked to observables, may be taken to exclude Bohm's theory from the range of theories permitted for the quantum domain.

In response, Bohm and Vigier maintained that the supporters of the Copenhagen interpretation are guilty of elevating the uncertainty principle into a principle of absolute and universal validity, and of employing it as a metaphysical principle to stifle further research at the sub-quantum level The true empirical approach is to judge the theory of hidden parameters on the basis of its success in explaining observations in the high-energy region, and not to judge the theory on a priori grounds.

Leon Rosenfeld was a leading member of the Copenhagen defense team. He replied that

> we must relinquish the illusion of being capable of unrolling a picture of the events on the Earth as they might be contemplated by some sharp-sighted, but not otherwise interfering, dweller of Sirius. . . . One of the lessons we have learned from the investigation of atomic phenomena is just that the quantum of action sets a limit to the unambiguous application of the mode of description of classical physics. To object to a lesson of experience by appealing to metaphysical preconceptions is unscientific.[15]

Option Three: Deny That Causality Holds in the Quantum Realm

It is incontrovertible that there are observed regularities among quantum phenomena. These regularities result from interactions of measuring apparatuses with quantum systems. The single-slit experiment is a good example. In this experiment there is an observed regularity between electron beams, of specified energy and direction, and impact patterns on photographic plates.

Such regularities do not establish causal relatedness, however. Causality (on option 3) is regularity between states of a physical system. The state of a quantum system is given by its Ψ-function, but this function contains an imaginary component. The empirical significance of the Ψ-function remains to be specified. If one accepts the standard Born interpretation, $|\Psi|^2 dV$ is the probability that the system is in volume element dV at a particular time.

To assign a state to a subatomic system, it is necessary to interact with the system. Aage Petersen noted that

> in any observational situation there must be some physical agency that is not included in the system and that serves to define the experimental conditions. Basic to the description of these conditions are the concepts of space and time.[16]

Decades earlier, Niels Bohr had insisted that

> any unambiguous use of the concepts of space and time refers to an experimental arrangement involving a transfer of momentum and energy, uncontrollable in principle, to fixed scales and synchronized clocks, which are required for the definition of the reference frame. Conversely, the account of phenomena which are characterized by the laws of conservation of momentum and energy involves in principle a renunciation of detailed space-time coordination.[17]

In option 3, causal relatedness is regularity between states of a physical system. The state of a physical system is well-defined only if the system is isolated from external influences. One lesson learned from investigations of quantum phenomena is that the values of state variables can be determined only by interacting with the system in question, and that the effects of this interaction cannot be eliminated or made negligible. A quantum system under investigation is not an isolated system. Thus the quantum system itself does not have a well-defined state.

There is a fundamental difference between classical physics and quantum physics in this regard. In classical physics, the interaction between the system under investigation and measuring instruments that assign values to state variables may be neglected or compensated for.[18] The isolated system is an ideal that may be approached asymptotically. In quantum physics, this is not the case. Every determination of the value of a state variable is subject to the restrictions placed by the uncertainty relations. The investigator has some latitude in deciding where "system" leaves off and "measuring apparatus" begins, but the presence of external influence on the system cannot be progressively eliminated. The object of study in the quantum realm invariably is the system under investigation. Bohr concluded that

> the finite interaction between object and measuring agencies conditioned by the very existence of the quantum of action entails—because of the impossibility of controlling the reaction of the object on the measuring instruments, if these are to serve their

> purpose—the necessity of a final renunciation of the classical ideal of causality.[19]

If Bohr is correct that one should renounce the classical ideal of causality in the quantum realm, then quantum physics presents no challenge to the "causality implies explanation" thesis.

Option Four: Appeal to Alternative Theories of Causality

A fourth option is to reevaluate the "causality implies explanation" thesis on some alternative theory about causal relatedness. Exchange of a conserved quantity, counterfactual dependence, and energy transfer are influential alternative positions on causality.

The Exchange of Conserved Quantities Interpretation of Causality

In the 1980s and 1990s, there was considerable interest in a view that interprets causality to be the exchange of conserved quantities, such as momentum, energy, charge, and spin. Since these quantities also are state variables in physical theories, the "exchange of conserved quantities" view is closely related to the "regularity-between-states" view.

The collision of billiard balls is causal because there is an exchange of a conserved quantity—momentum—in the collision process. In addition, if one takes position and momentum to be state variables for the system of cue ball and eight ball, then there is a causal regularity between successive states of the system. And of course, there is energy transfer in the process as well.

One distinctive feature of the conserved quantity view is Wesley Salmon's emphasis on a statistical relevance condition for causal relatedness. Salmon called attention to the case of an individual who is exposed to a low level of radiation.[20] Smith is situated a certain distance from a source of radiation, and he contracted leukemia. Only 2 percent of individuals at this distance from the source contract leukemia. The occurrence of leukemia is highly improbable. According to Salmon, Smith's exposure to radiation is the cause of his leukemia, despite the fact that he is a member of a small percentage of similarly situated persons to contract the disease. The decisive factor is statistical relevance and not mere statistical probability. What counts in Smith's case is the difference between the probability that Smith contracted leukemia and the probability that he contracted leukemia in the absence of the radiation source.

According to Salmon, a typical scientific explanation specifies a causal process that satisfies the statistical relevance condition. A causal process is the persistence of an entity, quality, or structure. Initially,

Salmon distinguished causal processes from pseudo-processes, such as the shadow of a moving car, by insisting that causal processes transmit marks impressed upon them, whereas pseudo-processes do not.

Several critics identified difficulties with the mark-transmission criterion. Salmon gave credit to Nancy Cartwright for the following criticism.[21] Suppose a rotating searchlight beam in the center of a circular room casts a moving circle of light around the room. The light path on the wall is a pseudo-process because pasting a red patch on the wall at one point in the beam's path does not affect the subsequent path of light on the wall. There is no mark transmission. However, if a red filter is placed over the searchlight lens at the same time that the red patch is placed on the wall, then there is mark transmission along the wall. On the mark-transmission criterion, the red path on the wall now is a causal process. To classify the moving red spot as a pseudo-process instead, one would have to appeal to the counterfactual claim that if the red filter had not been placed on the lens, then the red mark would not have been transmitted.

Salmon conceded that an appeal to counterfactual conditions is inappropriate in the application of a criterion. He subsequently abandoned mark transmission in favor of a criterion suggested by Phil Dowe, who maintained that "a causal process is a world-line of an object which manifests a conserved quantity."[22] He maintained also that a causal interaction is "an intersection of world-lines which involves an exchange of a conserved quantity."[23]

The collision between a ${}_7N^{14}$ atom and an α-particle qualifies as a causal interaction on the conserved quantity theory.

Figure 16
The ${}_7N^{14}$–α-Particle Collision

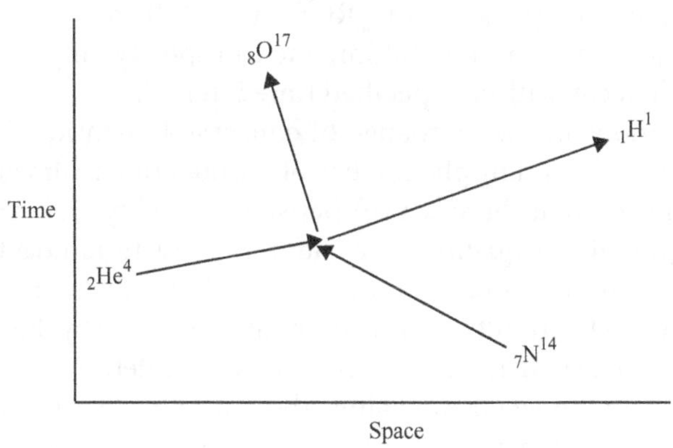

The conserved quantities mass-energy, momentum, and charge are exchanged in this collision process. Moreover, the probability of ${}_{8}O^{17}$ formation from ${}_{7}N^{14}$ is higher in the presence of α-particles than in their absence. The same quantities are conserved in spontaneous radioactive decay as well.

Figure 17
The Decay of ${}_{88}Ra^{226}$

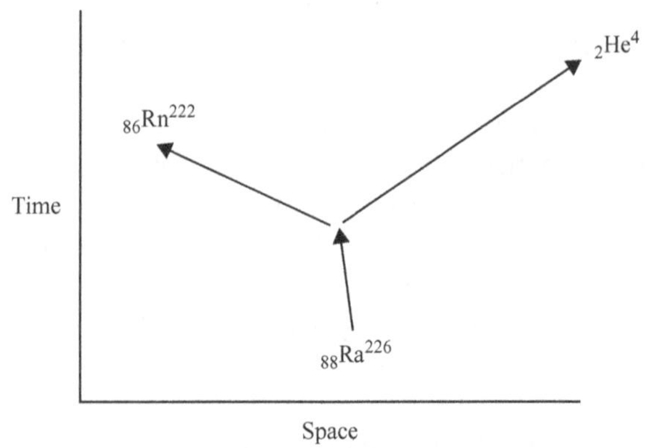

There is no interaction in this decay process. Nevertheless, Dowe held that instances of radioactive decay are causal processes, just as instances of inertial motion are causal processes.[24] This comparison is suspect. As indicated in chapter 3 (p. 36), in the case of inertial motion, there exist moving bodies that "manifest a conserved quantity." However, there is no object that persists throughout the process of radioactive decay. Moreover, although one can identify an "effect state" (the presence of ${}_{86}Rn^{222}$ and ${}_{2}He^{4}$) there is no obvious "cause state." Given a ${}_{88}Ra^{226}$ atom, one can specify only a probability that it will decay within a specified time interval.

Applications of the "exchange of conserved quantities" theory of causality to the puzzling phenomena of the quantum realm do not alter the conclusion that there is a rift between causality and explanation. Neutral pion decay qualifies as a causal process (if radioactive decay does), since spin is conserved throughout the process. However, no explanation of the results of measurement is provided by the conserved quantities theory. It remains a mystery why a determination of the value "spin up" for electron #1 alters the spin value of electron #2 from "50 percent up–50 percent down" (i.e., undetermined) to "spin down."

The two-slit experiment also qualifies as a causal relation on the "exchange of conserved quantities" theory. There is an exchange of conserved quantities energy and momentum with the diaphragm in that experimental arrangement in which it is free to move. In addition, there is an exchange of energy and momentum at the photographic plate. However, this causal account does not provide an explanation for the change produced in the pattern of impacts when electrons are detected at slit *A*.

The Contrary-to-Fact Conditional Interpretation of Causality

It has been suggested, by David Lewis and others, that causal relatedness is counterfactual dependence. To his constant sequential conjunction definition of cause, David Hume had appended the phrase

> or, in other words where, if the first object had not been, the second never had existed.[24]

Lewis recommended that people take "*c* causes *e*" to be logically equivalent to "if *c* had not occurred, *e* would not have occurred, given that *c* and *e* both do occur." In the usual notation

$$\text{“}c \text{ causes } e\text{”} \equiv (\sim c \;\Box\!\!\rightarrow \sim e)$$

where "$x \;\Box\!\!\rightarrow y$" means "if x had occurred, then y would have occurred." For a chain of causes,

"c causes e" $\equiv (\sim c \;\Box\!\!\rightarrow \sim x_1)$ & $(\sim x_1 \;\Box\!\!\rightarrow \sim x_2)$ & . . . $(\sim x_{n-1} \;\Box\!\!\rightarrow \sim x_n)$ & $(\sim x_n \Box\!\!\rightarrow \sim e)$.

The individual conditional terms are required to be "forward looking." "Backtracking" conditionals that claim that a past event would have been different if a present event were different are excluded. This restriction disqualifies counterfactual conditionals such as "if that child did not have blue eyes, then the parents would not have had blue eyes as children."

Clearly there is a problem about the truth status of proposed counterfactual conditional claims. In the real world, *c* and *e* both occur, but consider what would be the case in an alternative world in which *c* does not occur. Lewis suggested that

> a counterfactual is non-vacuously true if it takes less of a departure from reality to make the consequent true along with the antecedent than it does to make the antecedent true without the consequent.[25]

It is necessary then to compare the closeness of various hypothetical alternative worlds to the real world. Given that O (*c*) and O (*e*) stand for "*c* occurs" and "*e* occurs," [O (*c*) □→ O (*e*)] is true if, and only if, there is a world W^* in which O (*c*) and O (*e*) both are true, and W^* is closer to the actual world than any world in which O (*c*) is true and O (*e*) is false.

Lewis conceded that there is an ineliminable degree of vagueness to the concept "similar possible worlds." He maintained, however, that agreement of general laws is usually more important than agreement on matters of fact in assessing similarity.

If one takes causal relatedness to be counterfactual dependence, the spin determinations in the neutral pion experiment qualify as a causal relation. The measurement "up" for electron #1 caused the measurement of electron #2 to be "down" if, and only if, if the measurement of "up" for electron #1 were not the case, then the measurement of electron #2 to be "down" would not have been the case either. Two alternative worlds in which the antecedent is false are

W_1—no measurement is made of the spin direction of electron #1, and

W_2—the spin direction of electron #1 is measured and found to be "down."

In W_1, the antecedent is false, but the truth status of the consequent is undetermined. The probability that the measured value of the spin of electron #2 is "not down" is 50 percent. In W_2, the antecedent is false, and the consequent also is false.

It is intuitively clear that W_2 is closer than W_1 to the actual world (in which the spin direction of electron #1 is measured and found to be "up"). Consequently, since in W_2, the measured spin direction of electron #1 is "not up," and the measured spin direction of electron #2 is "not down," the measured spin directions of the two electrons qualifies as a causal relation. Since an explanation of this relation is not available (on pain of denying the special relativity prohibition on instantaneous influence over distance), the "causality implies explanation" thesis fails for neutral pion decay on the counterfactual conditional understanding of causality.

A similar analysis may be made of the two-slit experiment. The change of the distribution of electrons from an interference pattern to a superposition of two bell-shaped curves is presumably the effect of introducing an electron detector at slit *A*. This relation qualifies as causal on the regularity-between-states view. In the counterfactual conditional assessment, the closest relevant alternative world is a world in which electrons pass through slits *A* and *B* in the *absence* of a detector at slit *A*. In this world, the antecedent is false, and the consequent also is false. The real-world relation is causal on the contrary-to-fact conditional view of causality. The same disconnect between causality and explanation on the regularity-between-states view remains on the counterfactual dependence view.

Moreover, the counterfactual dependence view is subject to the criticism that it is neither a necessary condition nor a sufficient condition of causality. That counterfactual dependence is not necessary for causality is illustrated by cases of trumping.[26] The standard illustration of trumping is a battle situation in which a major and a sergeant simultaneously shout "Charge!" and the troops advance. If, contrary to fact, the major had not shouted "charge," the troops still would have advanced on the sergeant's command. The major's action is not a cause on the counterfactual dependence view, since the troops would have advanced even if he had not acted. The counterfactual conditional view qualifies as causal a case where either the major or the sergeant shout "charge," but it denies causal status to a case in which both the major and the sergeant shout "charge."

The counterfactual conditional interpretation is not a sufficient condition of causality either. There are empirically significant temporal sequences that satisfy the relation of counterfactual dependence but are not considered to be causal relations. Suppose that a melting point determination for an organic compound (e.g., aspirin) yields the sharp value *T* (e g., 134°C) at time t_1, and the compound reacts with sodium to form a new compound (e.g., sodium acetylsalicylate) of molecular weight *W* (e.g., 202.1) at time t_2. Melting-point determinations are effective in the identification of many organic compounds. If the measured melting-point had not been *T*, then a compound of molecular weight *W* would not have been formed with sodium. There is a relation of counterfactual dependence between the measured results. However, it would be implausible to claim that the melting-point determination is the cause of the formation of the sodium compound.

A similar analysis may be made of the relationship between the production of a green flame test by a substance (indicating the presence of the Ba^{++} ion) and its subsequent formation of a precipitate of molecular weight 233.1 with Na_2SO_4 (viz., $BaSO_4$). The production of $BaSO_4$ is counterfactually dependent on the green flame-test result. If no green color had been present in the flame, then no precipitate of molecular weight 233.1 would have been produced. However, the green flame-test result is not a cause of the subsequent precipitation of $BaSO_4$.

The Energy Transfer Interpretation of Causality

Energy transfer is a third alternative to the regularity-between-states interpretation of causality. This position has been championed by David Fair, among others. Fair maintained that scientists have come to understand that causal relatedness is energy/momentum transfer, just as they have come to understand that the temperature of a body is the mean kinetic energy of its molecules.[27] One might argue that our present understanding of causality within the context of scientific inquiry has evolved from (1) regularity between events, to (2) regularity between states of (relatively) isolated systems, to (3) energy/momentum transfer.

Fair noted that in most cases of causal interaction there is both energy transfer and momentum transfer. However, there are processes in which momentum transfer alone suffices to establish causal relatedness. The motion of a satellite around its planet is a case in point. The satellite is subject to continuous momentum change without a transfer of energy. Of course, gravitation is a mutually attractive force. Both bodies move relative to their common center of mass. The idea of motion subject to a $1/R^2$ force directed from a stationary central point is an idealization. Nevertheless, the satellite's motion, qua momentum change, is properly labeled an effect of the gravitational field of its planet.

In addition, Fair extended causal relatedness to processes in which potential energy is converted into kinetic energy. On this interpretation, bodies in free fall qualify as causal processes.

If causality is a transfer of energy or momentum, then the "causality implies explanation" thesis holds within quantum mechanics. In the neutral pion experiment, for instance, there is no transfer of energy from one electron to another, and hence no causal relatedness. The neutral pion experiment is not a challenge to the "causality implies explanation" thesis on this understanding of causality. However, there remains the puzzle about the seemingly instantaneous transmission of influence through space.

In the two-slit experiment, there is energy transfer between an electron and the diaphragm. This is a causal relation. There also is energy transfer from an electron to the photographic plate. This is a second causal relation. Explanations are available for each of these causal relations. The "causality implies explanation" thesis holds for these relations despite the fact that one cannot trace the path of an electron from diaphragm to photographic plate.

If the two-slit experiment is modified such that electrons passing through slit *A* are detected, the original diffraction pattern changes to a superposition of two bell-shaped distribution patterns. Since there is no energy transfer from the detector at slit *A* to electrons passing through slit *B*, no causal relation is present. The "causality implies explanation" thesis is not threatened.

Commitment to the energy transfer theory of causality does not remove perplexity however. Observation of electrons at slit *A* changes what happens at slit *B* in spite of the absence of energy transfer. It is puzzling to concede this, only to deny causal status to what happens at slit *A*.

If causal relatedness is energy transfer, then the "causality implies explanation" thesis may be retained in the face of the results of the neutral pion experiment and the two-slit experiments. The thesis may be retained only because there is no relevant energy transfer in either experiment. However, if causal relatedness is energy transfer, it would seem that one needs a quasi-causal category of influence at a distance in order to explain the results of the two experiments. Critics will complain that "influence at a distance" is a noncausal category and not a quasi-causal category, and that reference to a mysterious influence at a distance does not contribute to an explanation of these experimental results.

On behalf of the energy transfer theory of causality, one might argue that, because the connection between causality and explanation is very important, the theory of causality that best preserves this connection should be adopted. If it is desirable to retain the "causality implies explanation" thesis, then the energy transfer theory of causality is preferable to the regularity-between-states theory and the counterfactual dependence theory.

The identification of causal relatedness and energy transfer is not achieved without cost, however. People often speak of causality by omission (Gus's failure to water the rosebush caused its death) and causality by disconnection (Tom broke the circuit and caused the room to become dark). These normal types of causal discourse are noncausal on the energy transfer theory. This is perhaps not decisive against the energy transfer theory. People could train themselves to withhold the label "cause" from instances of omission and disconnection.

More importantly, the energy transfer theory qualifies as causal instances of radioactive decay, which instances surely are noncausal if any processes are. There is energy transfer from parent nucleus to daughter nucleus plus emission products, but, so far as is known, this energy transfer does not issue from an identifiable prior state of the parent nucleus. On the energy transfer theory, the spontaneous decay of a radioactive nucleus is a causal relation, the effect of which is clear, but the cause of which seems not to exist.

Option Five: Niels Bohr and the Limits of Causal Analysis

A fifth option is to follow the lead of Niels Bohr and examine the limitations placed on causal explanation by investigations of the quantum realm. Bohr declared that

> all knowledge presents itself within a conceptual framework adapted to account for previous experience and . . . any such frame may prove to be too narrow to comprehend new experience.[28]

Bohr sought to elaborate the epistemological lesson that he believed is implicit in the success of quantum theory.

Bohr took as starting point for this investigation the ability to communicate successfully what is known about causal interaction in everyday experience. The following is a prosaic example:

Georgia: "The lights in the living room were out for a minute, and I had to reset the electric clock."

George: "Sorry, it was my fault. I needed to shut off the power in the garage, and I threw the wrong switch."

Georgia: "Well, that is one outage we can't blame on Potomac Electric."

Communication about a causal relation has been achieved: Georgia understands that George caused the living room to become dark by engaging its circuit breaker. Presumably, Georgia and George share rudimentary background causal knowledge about electric circuits—namely, that illumination occurs only when there is a closed circuit through which electrons flow.

Bohr agreed with Wittgenstein that the meaning of a word is (usually) its use in a language.[29] He maintained that there are numerous examples of successful communication about causal relations that invoke, explicitly or implicitly, the principles of classical physics. According to Bohr, this must be taken as given. Edward MacKinnon emphasized that Bohr took the meaning of causality to be assigned by its use in everyday language, augmented by the resources of classical physics.[30]

Of course there are idealizations present in the conceptual structure of classical physics. A point mass, for instance, is not an entity encountered within experience. Inertial motion is likewise an empirically unrealizable abstraction. A body under the influence of no external forces would be a body whose motion could not be known, given universal gravitational attraction and the presence of other bodies necessary to measure constancy of motion.

Applications of the laws of classical physics presuppose a high level of abstraction. Nancy Cartwright has emphasized that to apply Coulomb's law to an interaction between charges is to presuppose that gravitational attraction may be neglected.[31] This presupposition holds strictly only for massless charges, which, like point masses, are not entities encountered within experience. Bohr held that the amalgam of everyday language and classical physics nevertheless satisfies the requirement of unambiguous communication.

Bohr notoriously maintained that he did not base his conclusions about epistemology on what philosophers have written. Nevertheless, his methodological approach resembles that of Immanuel Kant. This resemblance has been cited by John Honner, Edward MacKinnon, Henry J. Folse, and Dugald Murdoch, among others.[32]

Kant had posed the following question: Given the existence of synthetic a priori knowledge, what are the necessary conditions of the possibility of this knowledge? Bohr posed a similar question: Given the existence of the unambiguous communication of causal knowledge, what are the necessary conditions of the possibility of this communication?

Bohr's answer is that there must be instances of energy or momentum transfer that satisfy the conservation principles such that successive positions of the physical systems involved may be recorded accurately. This is to take the classical ideal of causality to be a condition necessary for the unambiguous communication of objective empirical knowledge. The classical ideal of causality includes both (1) the transfer of energy and/or momentum and (2) the concomitant spatiotemporal description of the physical systems involved in the transfer. Bohr discussed the regularity required by the classical ideal in terms of energy and/or momentum transfer subject to the relevant conservation principles, rather than in terms of a relation between states of a physical system. However, since he maintained that states of physical systems can be defined unambiguously in classical physics, statements about energy and/or momentum transfer can be translated into the terminology of the regularity-between-states view of causality.

The classical ideal of causality was applied consistently in the historical development of physical science in the eighteenth and nineteenth centuries. During this period, the history of science appears to conform, to a large extent, to the growth-by-incorporation model. William Whewell (1794–1866) likened the progress achieved within the history of science to the confluence of tributaries to form rivers.[33] He maintained that the history of theories about motion and light display a progressive incorporation of past results into present theories.

The transition from classical physics to quantum physics, by contrast, suggests a model of revolutionary overthrow, in which the successor theory emerges victorious after vanquishing its predecessor. Despite the appearance of revolution, Bohr maintained that there was an essential continuity in the transition from classical physics to quantum physics. He wrote that

> quantum mechanics presents a consistent generalization of deterministic mechanical description which it embraces as an asymptotic limit in the case of physical phenomena on a scale sufficiently large to allow the neglect of the quantum of action.[34]

The generalization of which Bohr speaks is achieved by means of an asymptotic agreement of calculations between quantum theory and classical theory for interactions involving massive bodies at relatively low velocities.

As previously noted, Bohr employed the correspondence principle as a methodological directive. In this usage, the correspondence principle requires that a theory of the subatomic domain yield calculations that agree with those of classical physics for that domain for which classical physics has achieved success.

Bohr also sought to extend the notion of correspondence to the use of concepts basic to classical physics in the new conceptual framework of quantum mechanics. Edward MacKinnon pointed out that Bohr's extension of the correspondence principle to concepts was undertaken in opposition to the operationalist position of Heisenberg. According to MacKinnon,

> Heisenberg insisted that the meaning of such crucial terms as "position" and "momentum" should be set by the operations through which these quantities are measured. Bohr insisted that such terms

> are not *assigned* meanings. Their meanings are already determined by their usage in classical physics.[35]

The same presumably is true of the concept causality. MacKinnon maintained that, for Bohr,

> the crucial conceptual problem was one of extending concepts, whose meanings are set in one conceptual framework, to a new conceptual framework resting on seemingly incompatible foundations. This problem was to be solved not by assigning meanings but by extending the correspondence principle from formulas to concepts.[36]

Heisenberg had defended the Kantian position that causality is a necessary condition of the possibility of objective empirical knowledge. Bohr agreed. He maintained that the unambiguous description of experimental results is a requirement for successful scientific inquiry at all levels. Bohr declared that

> experimental conditions can be varied in many ways, but the point is that in each case we must be able to communicate what we have done and what we have learned, and that therefore the functioning of the measuring instruments must be described within the framework of classical physical ideas.[37]

Application of the classical ideal of causality is a necessary condition of the unambiguous description of experimental results. For instance, to use an instrument to measure values of momentum or energy is to apply the principles of conservation of momentum or conservation of energy. The experimental apparatus must contain a movable part to which the conservation laws may be applied.

In classical mechanics, the concept of causal relatedness is applicable at the theoretical level as well as at the observational level. This is not the case in quantum mechanics, however. A central lesson of quantum mechanics is that the uncertainty relations place limitations on empirically meaningful applications of classical concepts. The classical ideal of causality cannot be applied unambiguously to the interpretation of the results of measurement in the quantum domain.

Bohr declared that

> every experimental arrangement permitting the registration of an atomic particle in a limited space-time domain demands fixed measuring rods and synchronized clocks which, from their very definition, excludes the control of momentum and energy transmitted to them.[38]

To locate a particle is to forego the opportunity to apply the conservation principles required for causal analysis. Bohr proceeded to note that

> conversely, any unambiguous application of the dynamical conservation laws in quantum physics requires that the description of the phenomenon involve a renunciation in principle of detailed space-time coordination.[39]

To provide a causal analysis by applying conservation principles is to forego the opportunity to trace precisely the spatiotemporal history of the physical systems involved. Applications of the concept of causality in quantum physics are limited to just one of two mutually exclusive experimental contexts. Bohr held that this mutual exclusivity of causal analysis and spatiotemporal description is a direct consequence of the uncertainty relations.

"State of a system" is a second concept whose use must be limited in the transition to the new conceptual framework of quantum theory. In classical physics, "state of a system" may be used in both a broad sense and a restricted sense.

In its broad sense, the state of a system is a set of values of state variables that can be predicated of the system whether or not these values are measured. Consider a body in free fall. Its velocity (a state variable) is measured after one second and found to be 32 ft./sec., and its velocity is also measured after three seconds and found to be 96 ft./sec. Physicists who apply the principles of classical mechanics have no hesitation in assigning "velocity is 64 ft./sec." to the body after two seconds of fall. They assume that this state was realized after two seconds, despite the absence of any experimental determination of this value.

In its restricted sense, the state of a system is a record of state variables measured at a specific time. When the concept "state of a system" is extended to the quantum domain, only the restricted sense is appropriate. One can communicate unambiguously about the state of a subatomic system, qua measured in a specific experimental arrangement. No specification of state variables between measurements is empirically significant.

Aage Petersen pointed out that "the possibility of isolating a physical system" (i.e., of eliminating all external influences) "is basic to the idea of a well-defined state of the system."[40] He added, however, that "the state concept is in turn basic to the idea of a strict connection between physical states" (i.e., to the concept of causality).[41] Since one cannot, as Bohr emphasized, neglect or compensate for, interactions between quantum systems and measuring instruments that assign values to state variables, one cannot gain information about an *isolated* quantum system.[42]

Bohr declared in "Theory of the Hydrogen Atom" (1927) that there is a relation of complementarity between spatiotemporal description and causality (qua regularity between states).[43] Spatiotemporal description may be achieved by interacting with the system through the manipulation of measuring instruments. An unambiguous specification of the state of the system is not possible in this context. One can, for instance, assign spatial and temporal values to an electron at a slit in a diaphragm and at a photographic plate. This assignment precludes a specification of the state of the system (independently of measurement) between observations. On the other hand, if there is unambiguous specification of the state of a system, the isolation required precludes spatiotemporal description.

It is not that there is no state function for a subatomic system between observations made of it. The Ψ-function represents its state during the intervals between measurements. However, this state is a probability value, and not a simple state variable, such as the velocity of a falling body. Since probability enters essentially into the definition of the state of a subatomic system, and does not enter into the definition of the state of a system in classical mechanics, the concept "state of a system" has undergone a significant change in the transition from classical physics to quantum physics.

One consequence of restricting in this way the use of the concept "state of a system" within quantum physics is to dissolve the puzzle over the reduction of the wave packet. Consider the diffraction of an electron at a single slit in a rigidly mounted diaphragm.

Figure 18
Diffraction at a Single Slit[44]

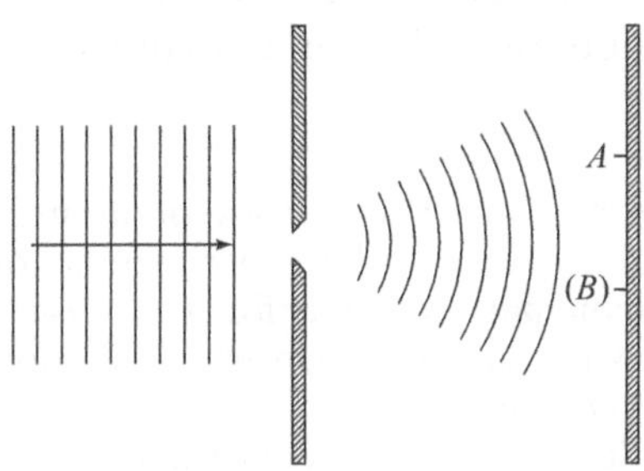

Referring to the above figure, Bohr declared that

> on account of the diffraction of the wave connected with the motion of the particle and indicated in the figure by the thin line, it is under such conditions not possible to predict with certainty at what point the electron will arrive at the photographic plate, but only to calculate the probability that, in an experiment, the electron will be found within any given region of the plate.[45]

Upon passing through the slit, presumably there is a spreading wave representing the probabilities that the electron strikes the plate at various points. Suppose the electron strikes the plate at specific point *A*. It would seem that the spreading wave has collapsed upon impact with the plate. The probability that the electron is at *A* is 1.

If the restricted version of "state of a system" is the only version permitted in quantum physics, then it is not meaningful to ascribe an objective state to the unobserved electron between slit and plate. The spreading wave and wave packet locutions are not part of the theoretical-level discourse of quantum theory.

Bohr reinstated these wave concepts in pictorial-level interpretations that are superimposed upon quantum systems between observations. He recommended the superposition of mutually exclusive, but complementary, wave and particle interpretations upon quantum systems between observations. However, these pictorial-level interpretations are "as if" interpretations. No claims are made about real waves spreading through space. It is Bohr's position that addition of these "as if" pictures is required by the ideal of intersubjective communicability.

The restricted applicability of the causal analysis of quantum phenomena is one lesson provided by quantum theory. A second lesson is that causal analysis nevertheless is required for the explanation of quantum phenomena. Bohr's position is not just that causal analysis is required for the determination of the results of measurement. He also insisted that causal analysis at the pictorial level is required for a complete account of the behavior of quantum systems. Bohr maintained that

> a most conspicuous characteristic of atomic physics is the novel relationship observed under experimental conditions demanding different elementary concepts for their description. Indeed, however contrasting such experiences might appear when attempting to picture a course of atomic processes on classical lines, they have to be considered as complementary in the sense that they represent

> equally essential knowledge about atomic systems and together exhaust this knowledge.[46]

The mutually exclusive, but complementary, pictures to which Bohr refers invoke causal analysis and spatiotemporal description. According to Bohr, these pictures combine to exhaust what can be known about the behavior of quantum systems between observations. He declared that

> evidence obtained under different experimental conditions cannot be comprehended within a single picture, but must be regarded as *complementary* in the sense that only the totality of the phenomena exhausts the possible information about the objects.[47]

Since a causal analysis is provided at the pictorial level for the behavior of quantum systems between observations, an explanation of this behavior is available. Of course, this explanation is incomplete, since it is not accompanied by a concomitant spatiotemporal description of the behavior of the system. The classical ideal of causality is inapplicable. A partial explanation is still an explanation, however. Bohr insisted that this partial explanation should be embraced as the best explanation available.

Notes

1. Philipp Frank, *Philosophy of Science* (Englewood Cliffs: Prentice-Hall, 1957), 264.
2. Ibid., 266–71.
3. David Bohm, *Causality and Chance in Modern Physics* (New York: Harper Torchbooks, 1957), 84.
4. Ibid., 95.
5. Ibid., 111.
6. Jean-Pierre Vigier, "The Concept of Probability in the Frame of the Probabilistic and Causal Interpretations of Quantum Mechanics" in *Observation and Interpretation*, ed. by S. Körner (New York: Academic Press, 1957), 75.
7. Bohm, *Causality and Chance in Modern Physics*, 80.
8. Ibid., 111–12.
9. Ibid., 111.
10. Ibid., 114.
11. Ibid., 115.
12. Ibid., 115–16.
13. Ibid., 115.
14. D. M. MacKay, "Comments," in *Observation and Interpretation*, 50–51.
15. Leon Rosenfeld, "Misunderstandings about the Foundations of Quantum Mechanics" in *Observation and Interpretation*, 44.

16. Aage Petersen, *Quantum Physics and the Philosophical Tradition* (Cambridge: MIT Press, 1968), 107.
17. Niels Bohr, "Unity of Knowledge" (1954), *Atomic Physics and Human Knowledge* (New York: John Wiley & Sons, 1958), 72.
18. Ibid., 72.
19. Ibid., 60.
20. Wesley Salmon, "Why Ask 'Why'? An Inquiry Concerning Scientific Explanation," *Proc. Am. Phil. Soc. 6* (1978), 689. Reprinted in *Scientific Knowledge*, ed. by J. Kourany (Belmont, CA: Wadsworth, 1987), 56.
21. Salmon, "Causality without Counterfactuals," *Phil. Sci. 61* (1994), 302–3.
22. Phil Dowe, *Physical Causation* (Cambridge: Cambridge University Press, 2000), 90.
23. Ibid.
24. David Hume, *An Enquiry* Concerning *Human Understanding* (Chicago: Open Court, 1927), 79.
25. David Lewis, "Causation," in *Causation*, ed. by Ernest Sosa and Michael Tooley (Oxford: Oxford University Press, 1993), 197.
26. Jonathan Schaffer, "Trumping Preemption," *J. Phil. 97* (2000), 175–76.
27. David Fair, "Causation and the Flow of Energy," *Erkenntnis 14* (1979), 220.
28. Bohr, "Unity of Knowledge" (1954), in *Atomic Physics and Human Knowledge*, 67.
29. Ludwig Wittgenstein, *Philosophical Investigations* (New York: MacMillan, 1960), Part I, §43, 20.
30. Edward MacKinnon, "Bohr on the Foundations of Quantum Theory," in *Niels Bohr: A Centenary Volume*, ed. by A. P. French and P. J. Kennedy (Cambridge: Harvard University Press, 1985), 115–16.
31. Nancy Cartwright, *How the Laws of Physics Lie* (Oxford: Oxford University Press, 1983), 57.
32. John Honner, *The Description of Nature* (Oxford: Clarendon Press, 1987), 200–201; 211–13;
    Edward MacKinnon, *Scientific Explanation and Atomic Physics* (Chicago: University of Chicago Press, 1982), 370–76;
    Henry J. Folse, *The Philosophy of Niels Bohr* (Amsterdam: North-Holland, 1985), 217–21;
    Dugald Murdoch, *Niels Bohr's Philosophy of Physics* (Cambridge: Cambridge University Press, 1987), 229–31.
33. William Whewell, *History of the Inductive Sciences* (New York: D. Appleton, 1859), 1:47.
34. Bohr, "Unity of Knowledge" (1954), in *Atomic Physics and Human Knowledge*, 73–74.
35. MacKinnon, "Bohr on the Foundations of Quantum Theory," 108.
36. Ibid., 108–9.
37. Bohr, "Atoms and Human Knowledge" (1955), in *Atomic Physics and Human Knowledge*, 39.
38. Ibid., 90.
39. Ibid.
40. Aage Petersen, *Quantum Physics and the Philosophical Tradition*, 107.
41. Ibid., 107.
42. Ibid., 116–18.

43. Bohr, "The Quantum Postulate and the Present Development of Atomic Theory" (1927), in *Atomic Physics and the Description of Nature* (Cambridge: Cambridge University Press, 1961), 54–55.
44. Bohr, "Discussion with Einstein . . ." (1949), in *Atomic Physics and Human Knowledge*, 42.
45. Ibid.
46. Bohr, "Unity of Knowledge" (1954), in *Atomic Physics and Human Knowledge*, 74.
47. Bohr, "Discussion with Einstein . . . ," in *Atomic Physics and Human Knowledge*, 40.

7

Complementarity in Biology

Bohr recommended complementarity as a criterion of completeness for explanations of quantum phenomena. To explain fully the single-slit diffraction of electrons, for instance, one must append two mutually exclusive pictures of the behavior of the electrons between the diaphragm and photographic plate. The two pictures are not applied arbitrarily. Spatiotemporal description (or the motion of classical waves) is appropriate given a rigidly mounted diaphragm. Causal interaction—via energy/momentum transfer (or the motion of classical particles)—is appropriate given a movable diaphragm.

Bohr held that these two pictures are appropriate because (1) the quantum of action places restrictions on the knowledge that can be achieved of a quantum system under observation, and (2) full explanation of a quantum process requires presentation in the common language used for unambiguous communication.

Bohr declared that

> the epistemological lesson of atomic physics has naturally given rise to renewed consideration of the use of our means of communication for objective description in other fields of knowledge.[1]

He suggested that the principle of complementarity is appropriate within disciplines other than physics. Bohr put forward the ideal of a "unity of science" in which different scientific disciplines share complementarity as a common methodological principle. Bohr discussed complementary interpretations in biology and psychology. An important question is how many aspects of complementarity, as a principle of atomic physics, are applicable within biological inquiry and psychological inquiry.

In the *Critique of Judgment* (1790), Immanuel Kant insisted that teleological interpretations of the form "*x* occurred in order that *y* be achieved" are required to supplement causal accounts of physiological processes. Whereas causal laws state that certain states of a living organism follow other states according to a rule, teleological interpretations provide answers to questions about the purpose served by observed patterns of structure, function, and behavior.

Kant maintained that teleological interpretations are required in biology because living organisms exhibit a reciprocal dependence of part and whole. An organism is what it is in virtue of the organization of its component organ systems. It also is the case that an organ system is what it is in virtue of its relations within the organism as a whole. Hence the need for teleological interpretations of the form "the function of *x* (e.g., the lungs) is to enable *y* (e.g., aeration of blood)." Kant emphasized that the inclusion of teleological interpretations contributes to the ideal of a systematic organization of biological knowledge.

Bohr stressed the need for teleological interpretations of activities, such as self-preservation and self-generation. For example, the human body maintains temperature homeostasis by means of a variety of adaptive responses to changes in the temperature of the environment. Among these adaptive responses are dilation and contraction of peripheral blood vessels, adrenaline gland secretions, the muscular contractions involved in shivering, and thyroid gland activity. A typical teleological interpretation is "Jones commenced to shiver, upon entering the frozen-food locker, so that his internal body temperature could remain (roughly) constant." In "Light and Life" (1932), Bohr called attention to

> the obvious exclusiveness between such typical aspects of life as the self-preservation and self-generation of individuals, on the one hand, and the subdivision necessary for any physical analysis on the other hand.[2]

The biologist who studies an animal's behavior in a particular ecosystem cannot perform the dissections and chemical analyses to uncover the underlying physiological changes within the animal without compromising the original study. Bohr maintained that

> every experimental arrangement with which we could study the behavior of the atoms constituting an organism . . . will exclude the possibility of maintaining the organism alive.[3]

Self-preservation and self-generation are processes fundamental to living organisms. Bohr maintained that the "notion of life" is an "element for which an explanation is neither possible nor required." In this respect, the "notion of life" plays a role in biology analogous to the role of the quantum of action *h* in quantum mechanics.[4] He held that the "existence of life" is a "basic postulate of biology" and, as such is "not susceptible to further analysis."[5] The life of an organism, like the stability of an atom, is a condition that is destroyed by observations made at a microlevel.

Bohr noted, moreover,

> the incessant exchange of matter which is inseparably connected with life will even imply the impossibility of regarding an organism as a well-defined system of material particles like the system considered in any account of the ordinary physical and chemical properties of matter.[6]

Werner Heisenberg provided a succinct summary of Bohr's position on complementarity in biology. In a reconstruction of a conservation with Bohr, Heisenberg reported that Bohr said:

> We can, first, describe an organism with concepts men have developed through contact with living beings over the millennia. In that case, we speak of "living," "organic function," "metabolism," "breathing," "healing," *etc*. Or else we can inquire into causal processes. Then we . . . study chemical or electrical processes, for instance in nerve conduction, and assume, apparently with great success, that the laws of physics and chemistry . . . are fully applicable to living organisms.[7]

According to Heisenberg, Bohr emphasized

> these two ways of looking at things are contradictory. For in the first case we assume that an event is determined by the purpose it serves, by its goal. In the second case we believe that an event is determined by its immediate predecessor.[8]

Bohr held that the two mutually exclusive approaches to biological inquiry are nevertheless complementary. Bohr was careful to distinguish his support for teleological interpretations in biology from the position that invokes the operation of vital forces to explain biological phenomena. According to Bohr, living organisms are complex, open systems properly characterized by causal analyses and teleological interpretations. There is no place in this biology for speculation about vital forces.

As applied to biological inquiry, the principle of complementarity requires, at a minimum, acceptance of mutually exclusive interpretations. Reports derived from experimental arrangements designed to measure and record adaptive responses of an organism in its environment, and concurrent reports derived from experimental arrangements designed to measure and record molecular reactions within the organism, are mutually exclusive.

The investigator of biological phenomena, like the investigator of subatomic phenomena, confronts mutually exclusive alternatives. One may study the behavior of an organism in its environment. To do so precludes conducting at the same time an investigation of molecular interactions within the organism. Alternatively, one may study interactions within the cells and organs systems of an organism. To do so precludes conducting concurrent studies of the adaptive responses of the organism in its environment.

What are mutually exclusive in biological investigations are selected experimental arrangements. One may set up an experimental arrangement either to observe physical and chemical interactions at the cellular level or to observe adaptive responses of an organism in its environment.

However, the resulting interpretations of physicochemical changes and behavioral responses are *not* mutually exclusive. They are interpretations given at two different levels, one of which requires application of teleological categories. By contrast, the particle pictures and the wave pictures put forward as interpretations of quantum phenomena are mutually exclusive. *Particle* and *wave* are antithetical concepts. A particle is an entity located, at a given time, within a limited region of space; a pure wave is non-localizable—it extends throughout all space. This important feature of complementarity in the Copenhagen interpretation of quantum mechanics does not carry over to biology.

Moreover, the mutually exclusive pictorial-level interpretations in terms of particles and waves recommended on the Copenhagen interpretation of quantum mechanics are superimposed upon a theoretical level of language within which references to classical particles and classical waves do not occur. There is no corresponding hierarchical linguistic usage in the interpretation of biological systems. Henry Folse observed that Bohr did not seem to recognize that

> the problem of keeping the organism alive while trying to observe it in a way necessary for a mechanical description of its component organs is, as he has analyzed it, apparently disanalogous to the quantum situation because the difficulty is merely a technical, practical difficulty and does not depend on the physical conditions required for unambiguously applying descriptive terms.[9]

What remains of the program to extend complementarity to biology is the mutual exclusivity of two types of experimental inquiry, both of which are required for the complete interpretation of biological phenomena.

Bohr held that an application of the principle of complementarity is required in quantum theory because the conceptual framework adequate to explain phenomena in classical physics proved too narrow to account for phenomena in the quantum realm. No such restriction of a previously effective conceptual framework is required to accommodate the use of the two mutually exclusive types of inquiry in biology.

The use of two mutually exclusive types of experimental inquiry in biology is not similar to the superposition of mutually exclusive wave and particle pictures to different experimental arrangements in quantum theory. Consequently, the evidence from biology provides no analogical support for the position that complementary wave and particle pictures are required for a complete explanation of quantum phenomena.

Notes

1. Niels Bohr, "Atoms and Human Knowledge" (1955), in *Atomic Physics and Human Knowledge* (New York: John Wiley & Sons, 1958), 91.
2. Bohr, "Light and Life" (1932), in *Atomic Physics and Human Knowledge*, 10.
3. Bohr, "Biology and Atomic Physics" (1937), in *Atomic Physics and Human Knowledge*, 20.
4. Bohr, "Unity of Knowledge" in *Atomic Physics and Human Knowledge*, 76.
5. Ibid., 21.
6. Ibid., 20–21.
7. Bohr, "Conversation with Heisenberg," in Werner Heisenberg, *Physics and Beyond* (New York: Harper Torchbooks, 1972) 110–11.
8. Ibid., 111.
9. Henry J. Folse, *The Philosophy of Niels Bohr* (Amsterdam: North-Holland, 1985), 189–90.

8

Complementarity in Psychology

Bohr accepted the distinction stressed by Kant and others between the awareness of an object and the awareness of being aware of that object. On one level, one may be directly aware of a Cezanne painting of Mount St. Victoire. On another level, one may catch oneself in a feeling of being exposed to beauty, or a feeling of being absorbed in a search for geometrical relationships in the painting, or a feeling of being involved in an attempt to recall a remembered image of the mountain itself.

Bohr noted that one of the "epistemological implications" of the success of quantum theory is the necessity of directing attention to the "placing of the subject-object distinction."[1] Any attempt to establish a sharp subject-object boundary within one's experience fails. The act of describing one's conscious awareness of a painting, for example, establishes the act as a further object of conscious awareness.

Every subjective state may be converted into an object for a higher level of subjective awareness. This shifting of the subject-object boundary introduces ambiguity into the interpretation of mental life. On one level, one may describe the interaction of a perceiving subject and a perceived object; on another level, one may describe the self-conscious awareness of the subject insofar as it is aware of its perceptual experience; and on a third level, one may describe the self-conscious awareness of the subject as it is aware of being aware of its perceptual experience.

Every attempt to capture a "pure subjectivity" converts it into an object. Nevertheless, Bohr maintained, the concept of "self-conscious awareness," like life in biological inquiry, is "indispensible when describing the human situation."[2] The experience of an object and the awareness of this experience are given together in one's conscious life. He observed that people become acutely aware of the indispensability of the concept of self-conscious awareness "in situations in which we ponder on the motives for our actions and in which we experience a feeling of volition."[3]

It is Bohr's position that a complete account of mental life must acknowledge continuous self-conscious awareness as a given in order to describe the volitional experience of the subject.

In a discussion with Heisenberg, Bohr declared that

> if we speak of free will, we refer to a situation in which we have to make decisions. This situation and the one in which we analyze the motives of our actions or even the one in which we study physiological processes, for instance the electrochemical processes in our brain, are mutually exclusive.[4]

Bohr maintained that the mutually exclusive causal and vitalistic interpretations of human behavior are complementary.

According to Bohr, a complete account of the life of the mind requires the incorporation of both causal and vitalistic points of view. He maintained that

> from a biological point of view, we can only interpret the characteristics of psychical phenomena by concluding that every conscious experience corresponds to a residual impression in the organism, amounting to an irreversible recording in the nervous system of the outcome of processes which are not open to introspection.[5]

The causal picture is one type of picture required for unambiguous communication about the life of the mind. In the complementary vitalistic picture, emphasis is placed on "the usefulness of permanent recordings in their influence on our reactions to subsequent stimuli."[6]

Bohr cautioned that to entertain vitalistic interpretations is not to support the position that there exist spiritual forces that influence matter. Vitalism is a perspective required within an empirical science of human behavior, spiritualism is not.

Unambiguous communication is an important goal of science. Ambiguities arise, however, in the uncritical interpretation of psychological processes. On one level, one can describe a subject's feeling of freedom as the subject chooses to perform a certain action. On another level, one can describe that feeling of freedom as the result of prior psychological states, themselves the result of physiological changes within changing environmental conditions. To move from the first level to the second, as if the same object were being described, is to create ambiguity. On the first level, the description is of the subjective awareness of a perduring consciousness. On the second level, the description is of an object situated within a causal chain. The subject-object boundary is movable, and transitions between the levels shift this boundary.

It is because there is a movable subject-object boundary in psychological inquiry, which is analogous to the moving system-instrument boundary in atomic physics, that complementary descriptions are appropriate in psychology as well. Bohr held that, in general, complementary descriptions are required to represent phenomena, the realization of which depends on the conditions under which they are obtained.

Bohr declared that

> when considering the contrast between the feeling of free will, which governs the psychic life, and the apparently uninterrupted causal chain of the accompanying physiological processes, the thought has, indeed, not eluded philosophers that we may be concerned here with an unvisualizable relation of complementarity.[7]

Bohr maintained that his recommendation is consistent with common linguistic usage. In ordinary usage, one can employ descriptions at both levels as if they complemented each other. Bohr concluded that a complete interpretation of human behavior includes both introspective reports from human agents and empirical investigations of agent-environment interactions.

Causal interpretations and vitalistic interpretations are held to be complementary. How effective is the analogy Bohr draws to complementary descriptions in atomic physics? In atomic physics what is investigated is not a quantum system, per se, but a quantum system-observed-by-means-of-a-specific-experimental-arrangement. On the Copenhagen interpretation, one describes a quantum system between observations by selecting either a particle picture or a wave picture. These two pictures are mutually exclusive. Each is appropriate for a specific experimental arrangement.

Causal pictures and vitalistic pictures, like particle pictures and wave pictures in quantum theory, are interpretations that supplement each other. These pictures are complementary in the sense that both types of picture are required for completeness. This is the case both in quantum physics and psychology.

Moreover, the choice of the appropriate picture is determined by the experimental arrangement selected, in psychology as well as quantum physics. If one selects introspection, then the vitalistic picture is appropriate. If one selects physiological inquiry, then the causal picture is appropriate.

However, causal and vitalistic pictures are not "mutually exclusive" in the way in which particle pictures and wave pictures are. Particle and wave pictures are "mutually exclusive" in a strong sense. A wave-particle is a self-contradictory concept. An electron, for example, displays wave-like properties, when observed under certain experimental arrangements, and particle-like properties, when observed under other experimental arrangements. But no entity is simultaneously both wave and particle. Such an entity would be both localized and infinitely extended. On the Copenhagen interpretation of quantum mechanics there is a prohibition against specifying both a wave picture and a particle picture for a subatomic system observed under a specific experimental arrangement.

Causal interpretations and vitalistic interpretations are mutually exclusive only in the weak sense that an introspective inquiry does not yield physiological data, and physiological inquiry does not yield information about the stream of self-conscious awareness. There is no prohibition against applying both causal and vitalistic interpretations to the same instance of behavior.

For instance, professional pickpocket Paul spies a bulging wallet in Wally's unbuttoned back pocket. He knows that, as a two-time loser, he runs a considerable risk of extended incarceration if he is caught in the act of pilfering the wallet. Paul is aware of deliberating about the situation and then acting to steal the wallet nevertheless. This introspective account is not inconsistent with a causal interpretation that accounts for his theft by reference to a pattern of behavior in the past.

The mutual exclusivity of interpretations is a cornerstone of the principle of complementarity, as it is applied within quantum theory. Bohr's analogical extension of the principle of complementarity to psychology is unconvincing. A more appropriate analogy to causal and vitalistic interpretations in psychology would be from fluid and particulate models of the behavior of aqueous solutions.

Notes

1. Bohr, "Unity of knowledge" (1954), in *Atomic Physics and Human Knowledge* (New York: John Wiley & Sons, 1958), 78–79.
2. Bohr, "Atoms and Human Knowledge" (1955), in *Atomic Physics and Human Knowledge*, 92–93.

3. Bohr, "Unity of Knowledge" (1954), in *Atomic Physics and Human Knowledge*, 77.
4. Bohr, "Conversation with Heisenberg," in Werner Heisenberg, *Physics and Beyond* (New York: Harper Torchbooks, 1971), 91.
5. Bohr, "Unity of Knowledge," in *Atomic Physics and Human* Knowledge, 77.
6. Ibid.
7. Bohr, "The Quantum of Action and the Description of Nature," in *Atomic Physics and the Description of Nature* (Cambridge: Cambridge University Press, 1961), 100.

9

Complementarity in Religious Language

There are occasional references to religion in Bohr's writings, but no sustained analysis of theology or religious discourse. Abraham Pais observed that Bohr "could appreciate religious feelings but not what one might call its philosophy."[1]

The following analysis is restricted to those theological interpretations that make putatively true claims about God and the world. No attention is given to theological presentations for which no support is provided over and above "it says so in such and such passage of scripture."

Proclamations about the relation between God and man in Christian theology interpret God to be either an immutable being or a partner in a dialog, so this study will refer to the two types of claim as being framed within a "language of immutable attributes" and a "language of dialog." Strictly speaking, of course, these are two modes of expression within a single language.

These two modes of expression are mutually exclusive. A dialog is genuine only if it is possible that each participant change in the context of interaction. Within the language of dialog, God solicits obedience, issues warnings, renders praise, and promises future benefits. These are aspects of a *personal* encounter, characterized by growth and development achieved by means of striving and self-overcoming. On the other hand, immutable attributes (e.g., omnipotence, omniscience, infinite benevolence . . .) are not subject to change. It would be a self-contradiction to state that an omnipotent being lost power or that an omniscient being gained knowledge.

There is, therefore, a similarity between discourse about particle and wave pictures in the Copenhagen version of quantum theory and discourse about immutable attributes and dialog in Christian theology. Bohr maintained that superposition of mutually exclusive particle and wave pictures is essential for a complete explanation of quantum phenomena. It is open to the theologian to claim that mutually exclusive pictures about immutability and dialog likewise are essential for theological explanation. By so doing, the theologian might claim support for this position by insisting that, since mutually exclusive pictures are complementary in the successful discipline of quantum mechanics, mutually exclusive pictures may be complementary in theology as well.

The following brief survey should suffice to establish that the language of immutable attributes and the language of dialog have been employed together by important Christian theologians.

Application of the Two Languages by Christian Theologians

St. Augustine (354–430)

The Confessions of St. Augustine take the form of an extended conversation with God.[2] Augustine makes use of the language of dialog throughout. He addresses God in the conviction that God understands and responds. Although Augustine may not have realized this, to be a partner in a genuine dialog requires a potentiality for change, growth, and development. Without the possibility of change, a supposed dialog is a shadow play.

The *Confessions* record Augustine's recognition of changes in his self-understanding as partner in a dialog with God. He came to realize that the essence of his dialogic relationship with God is an orientation of the will. The proper orientation is a self-denying love of God, in which he acknowledges his complete dependence on the God who sustains and illuminates him. The improper orientation is a God-denying love of self, in which he elects an illusory self-sufficiency.

Amor dei (the self-denying love of God) and *amor sui* (the God-denying love of self) also serve to delineate communities on the stage of world history. The "city of God" is populated by men who orient their respective wills to God and express their orientation through charitable concern for their fellow men. The "city of man" is populated by men whose elevation of self-interest is expressed through strategies designed to secure dominance over others.

The *City of God* traces the interaction of the two cities throughout history.[3] The city of God has its nucleus in the church. The city of man has its nucleus in the empire. Augustine interpreted developments in world history by reference to an ongoing dialog between God and his chosen people. Within the dialog, God is interpreted as issuing commands and warnings, as well as promising rewards for proper behavior.

Although Augustine's primary theological usage invokes the language of dialog, he also made use of the language of immutable attributes when the occasion demanded it. He opposed the Manichean heresy that claimed that there is a battle for men's souls between a "Prince of Light" and a "Prince of Darkness" by insisting that there is but one extra-human Prince, and that this God is omniscient, omnipotent, and infinitely benevolent.[4] There is no Prince of Darkness. Evil is a mere lack and not some positive agency. Augustine agreed on this point with the neo-Platonic philosopher Plotinus.

God is omnipotent. Augustine noted that God could have created man such that man could not sin (*non posse peccare*). That He did not do so must mean that as an infinitely benevolent being, God's decision to create Adam with free will (*posse peccare et posse non peccare*) must represent a greater perfection because it paves the way for the redemption of sinful man through Christ and his church.

St. Thomas Aquinas (1225–1274)

Augustine chose the language of dialog as the principal vehicle in which to express his theology. St. Thomas Aquinas, by contrast, placed greater emphasis on the language of immutable attributes. Thomas maintained that human reason is competent to prove certain truths about the existence and attributes of God. These truths are contained in five arguments.[5] The arguments are deductive in form, with premises about the world and conclusions about the nature of God. The arguments thus proceed from claims about existence to claims about essence. This is not the place to argue the case, but no one of these arguments is sound.

If the five arguments were sound, Thomas would have demonstrated that there is a necessarily existent, omniscient, most perfect being, the unmoved originator of all motion and the uncaused producer of all effects in the universe. Thomas believed that the five arguments provide a persuasive case to change the minds of those who entertain doubts about the truths of the Christian faith. Thomas maintained that, in addition to the truths of natural theology that can be established by human reason, there are additional truths revealed to man by God. These revealed truths include original sin, the Incarnation, the resurrection of the body, and the sacramental system that forms the basis for activity within the church.

Thomas superimposed upon his discussion of the immutable nature of God a dialogic account of the relation between Christ and man. This account includes depicting Christ as a dialog partner who sets forth moral precepts and illustrates their application by means of his own life. Christ is both exemplar and teacher to man. To express the significance of his life to the faithful, the theologian must utilize a language within which growth and development takes place.

Martin Luther (1483–1546)

Martin Luther emphasized the proclamations and activities of a personal god, a god with whom one enters into a dialog. In this dialog, God freely issues grace to man, thereby creating a relationship of faith. Luther held that those in whom faith has been created are justified by this faith alone.

Grace is a manifestation of God's love. Luther called attention to various manifestations of God's love, which are revealed in scripture, and to manifestations of God's wrath as well. This exposition is set within the language of dialog.

The manifestations of love and wrath do not alter the nature of God. Luther was quite clear that God's knowledge, power, justice, and goodness are not qualities subject to change. Addressing Erasmus, Luther declared that

> Do you suppose that He does not will what He foreknows, or that He does not foreknow what He wills? If He wills what He foreknows, His will is eternal and changeless, because His nature is so.[6]

Our dialogic relationship is, paradoxically, a relation within which the other partner does not, and cannot, change.

Luther distinguished between the "Word of God" and "God in His own nature and majesty."[7] The scriptures reveal God's Word, a communication designed to guide the behavior of men. It is a narrative presented in the language of dialog. The narrative depicts divine promises, commands, and actions in which possibilities become realized. God in his own nature and majesty, however, is immutable.

Soren Kierkegaard (1813–1855)

Soren Kierkegaard emphasized that the Christian religion is based on an "Absolute Paradox."[8] The paradox is that God, an immutable being, has elected to enter human history, thereby actualizing previously unrealized possibilities. This transformation occurred at a decisive point in time. It is an event that is inconceivable from the human point of view. One cannot conceive how an immutable being, not subject to change, can become a being open to possibilities of growth and development while remaining an immutable being.

Kierkegaard applied both the language of dialog and the language of immutable attributes in his expression of the Absolute Paradox. He found that the language of dialog is appropriate to characterize the relationship between Christ and man. Christ is the teacher who confronts and engages the self. The self, qua sinner, has sought to constitute itself as a self-sufficient entity. It pursues a self-affirmation that denies its fundamental dependence on the power that created it. Sin is a defiant misorientation of the will.[9]

According to Kierkegaard, Socrates exemplifies the highest level of instruction on a purely human level. Socrates practices *maieusis* (midwifery), drawing forth from his students the latent knowledge they already possessed. Christ, as teacher, is not a midwife. He is a begetter, who creates a new relation of faith in which the entire orientation of the self is redirected. The teacher not only brings truth to the self but also brings the conditions required for its appropriation. Truth is established in a direct personal relationship in which he learner achieves "contemporaneity with Christ." Kierkegaard was preoccupied with this one-on-one relationship and believed that institutional expressions of religion are more likely than not to impede its development.

It is an absolute paradox that the teacher who engages in dialog with a sinful man is an immutable, omniscient, and omnipotent being. This absolute paradox is an affront to human reason. Kierkegaard insisted that man must turn away from the conclusions of reason in order to embrace the Absolute Paradox. The truth that matters, he notoriously declared, is "an objective uncertainty held fast in an appropriation-process of the most passionate inwardness."[10]

The Two Languages: Mutually Exclusive but Complementary?

The above survey is perhaps enough to indicate that representative Christian theologies are expressed in both a language of dialog and a language of immutable attributes. This usage is present as well in typical discussions between believer and skeptic:

> *B*—"God responds to our prayers because he loves us."
>
> *S*—"In learning of our desires and needs and deciding to respond favorably, does God gain in knowledge or goodness? Surely

> if God listens and responds, he is a being subject to growth and development."
>
> *B*—"Don't be naïve. God is omniscient, omnipotent, and infinitely benevolent. He does not grow in knowledge, power, or goodness as a result of his interactions with us."

In the novel *1984*, George Orwell introduced the strategy of "doublethink."[11] The individual who practices doublethink applies logically incompatible interpretations in alternative contexts. In a Christian religious context, the logically incompatible interpretations are those expressed in the language of dialog and the language of immutable attributes. The practitioner of Orwellian doublethink shifts from one language to another as the occasion demands.

The problem for those who practice Orwellian doublethink is that any conclusion can be derived from inconsistent premises. Given that *A* and *not A* both are true, as practitioners of doublethink maintain, it follows that *X* and *not X* both are true, regardless of what *X* states:

1.	A	given	A
2.	$\sim A$	given	$\sim A$
3.	$A \vee X$	1, Add.	$A \vee \sim X$
	$\therefore X$	2, 3, D.S.	$\therefore \sim X$

For instance, if "God is immutable" and "God is not immutable" both are true, then "the earth is a sphere" and "the earth is not a sphere" both are true as well. A theology derived from inconsistent premises is not cognitively significant.

There are at least three plausible responses to the Orwellian challenge to theology. One approach is to affirm the Absolute Paradox and base all claims on a faith not subject to rational analysis. The skeptic will protest, of course, that a self-contradiction is a self-contradiction, and that to refer to it as an "Absolute Paradox" contributes nothing to the task of theology.

A second approach is to restate theological interpretations either exclusively in the language of immutable attributes or exclusively in the language of dialog. Several thinkers have selected this approach. The philosopher Baruch Spinoza (1632–77) formulated a theological system without recourse to the language of dialog. Spinoza selected axioms that predicate of God such attributes as "infinite," "eternal," "indivisible," "necessarily existent," and "first cause."[12] He then derived from these axioms statements about God's immanent presence within the universe. Critics have debated whether Spinoza's deductive system qualifies as a theology. It clearly is not an expression of Christian theology.

The theologian Edwin Lewis (1881–1959), by contrast, formulated a version of Christian theology without recourse to the language of immutable attributes. Lewis's God is limited in power and is subject to growth. His power as Creator is opposed by an Adversary.[13] Just as Plato's Demiurgos is limited in his imposition of forms by the form-resistant power of the Receptacle, so also the Creator is limited by the power of the Adversary. But the Adversary, unlike Plato's Receptacle, possesses a positive disruptive power. The Adversary is a nonbeing in the sense of *me on*, an active power that nihilates the structures of being, and not a nonbeing in the sense of *ouk on*, a mere privation or absence of being.

According to Lewis, the Creator has revealed himself to be a personal being. God has motives and makes choices. He has chosen to participate in a dialog with men, a dialog whose course is revealed in the scriptures. The resurrection of Jesus Christ is the decisive moment in the struggle against the Adversary. It marks the final triumph of the Creator over the Adversary.

Lewis's theology contains a substantial whiff of Manichean dualism. As such, it is heretical from the standpoint of traditional Christian theology. However, it is not subject to the challenge of a conflation of inconsistent predications of dialogic categories and immutable attributes.

A third approach is to argue that interpretations in the language of dialog and interpretations in the language of immutable attributes are complementary in the same way that "moving particle" accounts and "spreading wave" accounts are complementary in the Copenhagen interpretation of quantum mechanics. In this approach, mutually exclusive, but complementary, dialogic accounts and immutability accounts are both necessary for a complete explanation of the relation between God and man.

In quantum physics, the claim that "that entity is both a particle and a wave" is a self-contradiction. Self-contradictory claims are not cognitively significant. However, the claim that in contexts of type *A* that entity behaves as if it were a particle, whereas in contexts of type *B* that entity behaves as if it were a wave is cognitively significant. Perhaps a similar comparison is relevant for theological claims about dialog and immutable attributes.

A number of conditions must be fulfilled for a convincing analogical extrapolation of complementarity to religious discourse. Among these conditions are:

1. The theology under consideration has a hierarchical linguistic structure that mirrors the structure of the Copenhagen interpretation;
2. Interpretations in the language of dialog and interpretations in the language of immutable attributes are mutually exclusive pictorial-level additions to a theoretical level of discourse; and
3. Each mutually exclusive pictorial-level interpretation is appropriate for a specific type of observational-level context.

Paul Tillich's Theology of Correlation

Paul Tillich's theology is a strong candidate to fulfill the above three conditions. Tillich sought to correlate existential questions that arise within a particular cultural situation with theological answers derived from his interpretation of the Christian message.[14] His theology appears to be expressed in a hierarchy of language levels that replicates the structure of the Copenhagen interpretation.

Language Level	Copenhagen Interpretation	Tillich's Theology of Correlation
Observational	Descriptions of experimental results in the concepts of classical physics	Existential questions that arise within a cultural situation
Theoretical	$d\Psi / dt$ on the Born interpretation of the Ψ-function	Ontological structure of essential being revealed in the New Being of Jesus as the Christ
Pictorial	Mutually exclusive causal interactions and spatiotemporal descriptions (or classical waves and particles)	Mutually exclusive descriptions in terms of dialog and immutable attributes

At the observational level, existential questions arise in subjective states of high intensity in which the self is overwhelmed by anxiety about loss of meaning, guilt, or death. The following questions are representative of this existential anxiety:

1. Are my achievements in life of any consequence?
2. Why am I so self-centered that I continually fail to respond to the needs of others?

3. What impact can I, as an individual, have on an impersonal, money-directed political process?
4. Why must I die?

These existential questions are expressed independently of reference to theological categories, just as the description of experimental arrangements and results are expressed independently of reference to the concepts of quantum physics.

At the theoretical level, Tillich postulated a dialectical relationship between the ontological categories *being* and *nonbeing*. The relationship is dialectical in the sense that the component realities are interdependent, such that any attempt to characterize one reality without reference to the other is unsuccessful. The term being, he declared, "means the whole of human reality, the structure, meaning, and aim of existence."[15] The structure, meaning, and aim of human existence are continually under threat from nonbeing, a power that possesses the capacity to nihilate these aspects of being. Being is ontologically prior to nonbeing. Negation is parasitic upon that which it negates. Nevertheless, nonbeing is a genuine threat to being.

Tillich referred to the tension between being and nonbeing as the state of "existential being." He maintained that this state constitutes a "fall" from the state of "essential being," in which estrangement is not present. In the state of essential being, being is supported by a relationship to "Being-itself."

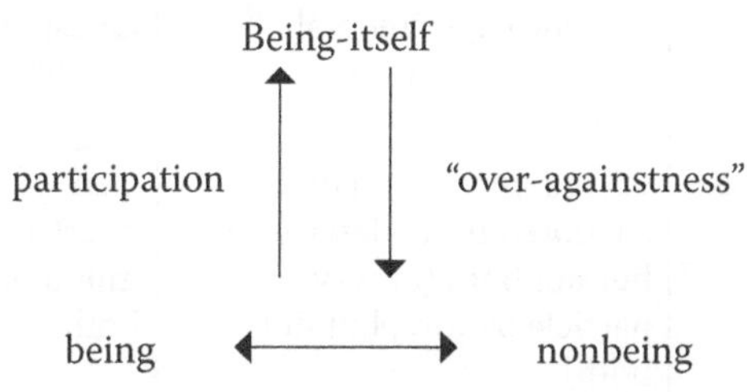

Tillich identified Being-itself and God, thereby interpreting "participation" and "over-againstness" as God's immanent (qua Holy Spirit) and transcendent (qua Creator) relation to the world.

The state of essential being is an ideal to which existential being is compared. Tillich maintained, however, that this ideal state has been realized under the conditions of human existence. This realization took place in the New Being in Jesus as the Christ.[16] The New Being is the theological answer to questions that arise in the state of existential estrangement. It reveals to man the unconditional power that lies behind and sustains the dialectical relation between being and nonbeing. The power of the New Being makes possible a self-affirmation *in spite of* the threat of nihilation from nonbeing.

The New Being is a higher ontological level than existential being because it is a state in which estrangement is overcome. It also is a higher ontological level than essential being because it is an *actualization* of the ideal of essential being under the conditions of human existence.

The theoretical-level response to questions that arise within the context of existential estrangement invokes a theory about ontological levels. This will strike most religious persons as overly abstract. To render theological answers to existential questions less abstract, Tillich incorporated into his theology a pictorial level of interpretation. At the pictorial level, mutually exclusive accounts in the language of immutable attributes and the language of dialog are superimposed on the ontological structure of the theoretical level.

Tillich did not support his pictorial-level accounts by reference to a similar use of mutually exclusive pictures in the Copenhagen interpretation of quantum mechanics. An enthusiast for his theology of correlation might do so however.

	Copenhagen Interpretation	**Theology of Correlation**
Subject of Interpretation	Specific experimental arrangement to study a quantum process	Questions arising from a state of existential estrangement
Mutually Exclusive Pictures	Causal interaction or spatiotemporal description, but not both (or wave or particle picture, but not both)	Dialogic interaction or attribution of immutability, but not both
Complementarity	There exists another experimental arrangement for the process for which the other picture is appropriate	There exists another aspect of the state of estrangement for which the other picture is appropriate
Completeness	Both pictures required for a complete explanation of a quantum process	Both pictures required for a compete explanation of the human condition

Pictorial-level theological accounts are subject to two requirements. They must be consistent with theoretical-level assertions, and they must be effective as answers to questions that arise in the state of existential estrangement. To be effective, the choice of a picture must be appropriate to the context to which it is applied.

In quantum mechanics, the selection of an experimental arrangement dictates which pictorial-level interpretation is appropriate. Does the existential context likewise dictate whether a dialogic account or an account in terms of immutable attributes is appropriate?

In *The Courage To Be*, Tillich delineated a three-fold threat posed by nonbeing. Nonbeing threatens man's self-affirmation ontically, spiritually, and morally.[17] Man's ontic self-affirmation is under a threat both from a fate that he does not control, and ultimately from the final nihilation of death. Man's spiritual self-affirmation is under a threat that denies significance to his personal achievements and the network of his interpersonal relationships. Man's moral self-affirmation is under a threat that exposes the self-seeking that masks his efforts to achieve good.

It would strengthen the analogical extrapolation of complementarity to theology if pictorial-level references to the immutable attributes of God are correlated only with existential questions that reflect ontic anxiety, and pictorial-level dialogic accounts are correlated only with questions that reflect spiritual anxiety or moral anxiety. This correlation of pictorial-level interpretations with different types of existential questions is plausible at first glance. Anxiety about one's impending death may be alleviated by assurance that an omnipotent God is able to set aside the power of nonbeing and that death is not the end. Anxiety about the loss of meaning or moral standing may be alleviated by assurance that a Personal Being exists, who cares about, and accepts, one as a person.

However, Tillich required only that a pictorial-level interpretation be effective. It must be accepted, and acted upon, by the individual who posed the existential question. An effective theological answer performs a quasi-therapeutic function. No restriction is placed on the type of picture to be employed in a given existential context. An account in the language of dialog may be effective in cases of ontic anxiety, and an account in the language of immutable attributes may be effective in cases of spiritual or moral anxiety. Tillich's position is unassailable. However, it establishes an important disanalogy to the use of mutually exclusive pictures in the Copenhagen interpretation of quantum mechanics. In the quantum realm, the nature of the quantum-system-under-observation determines the type of complementary picture to be applied.

Tillich did not present his theology in the form of a deductive system. Nevertheless, he did affirm a set of inviolable principles that have axiomatic status within his theology. These principles are

1. There is a triadic relationship of being, nonbeing, and Being-itself.[18]
2. God is Being-itself.[19]
3. Personal Being is the preeminent symbol appropriate to God.[20]
4. All statements about God, except (2) and (3) are symbolic claims.[21]
5. Existential estrangement is overcome in the appearance of essential being under the conditions of human existence in the New Being in Jesus as the Christ.[22]

Given these basic principles, theological answers to existential questions rely on the power of symbols. Tillich maintained that an effective symbol both points beyond itself and participates in the reality of that toward which it points. Examples of symbols that have proved effective include the cross, flags, anthems, and paintings. How can one know whether a particular symbol is effective in a given context? One can cite examples. Picasso's *Guernica*, for instance, has brought home to many twentieth-century viewers the horrors of war. Battle scenes by Uccello have not displayed this power. One cannot, however, specify necessary or sufficient conditions for the effectiveness of a symbol. A theological answer expressed in symbolic images of a personal god is either effective in a given context or it is not. Tillich maintained that

> the New Being has the power to transform those who are transformed by it.[23]

A pictorial-level answer is effective if, and only if, it is accepted and acted upon by those who initially posed the existential question.

However, if that is the case, then the analogical extrapolation of complementarity to religious language is a failure. The Copenhagen interpretation appends "as if" pictures to quantum theory. Given a beam of electrons incident upon a slit in a rigidly mounted diaphragm, it is "as if" a classical wave spreads from slit to photographic plate. No claim is made that there really exists a classical wave between slit and plate. Nevertheless, Bohr and supporters of the Copenhagen interpretation claim explanatory significance for "as if" pictorial-level accounts.

Support for this claim of explanatory significance can be found in other domains of science. The ideal gas theory, for instance, pictures gases as if they were a collection of elastically colliding point masses. The theory has explanatory power despite the fact that no claim is made that such entities exist. Of course, a theory may have explanatory power in a domain, even though there is another theory in that domain with greater explanatory power.

One might suggest that, since "as if" interpretations often are taken to be explanatory in science, in spite of the fact that no claims are made that point masses, massless charges, or ideal pendulums exist, theological interpretations that invoke "as if" assertions also are explanatory. This comparison would be inappropriate. Scientific theories that make "as if" claims about point masses and ideal gases are indeed explanatory, but they are not explanatory *because* they make no claims about existence. The same is true of "as if" theological interpretations. They do not gain explanatory status *because* they make no claims about existence. The explanatory significance of theological interpretations must be established on other grounds.

The analogical extrapolation of complementarity to theology applies mutually exclusive interpretations in the language of immutable attributes and the language of dialog. On the complementarity analogy, these interpretations also are "as if" interpretations. However, "as if" interpretations are inadequate as responses to existential questions. To be told that "it is as if you are in a dialog with a Personal Being" is not likely to reassure an individual anxious about impending death, loss of meaning, or perceived guilt. Despite its initial plausibility, the analogical extrapolation of complementarity to assertions about immutable attributes and assertions about dialog is not an effective defense against the charge of Orwellian doublethink.

Notes

1. Abraham Pais, *Niels Bohr's Times* (Oxford: Clarendon Press, 1991), 24.
2. St. Augustine, *The Confessions of Saint Augustine* (New York: Random House, 1949).
3. St. Augustine, *The City of God* (New York: Random House, 1950), books XI–XVIII.
4. St. Augustine, *The Nature of the Good, Against the Manicheans* in *Augustine: Earlier Writings*, ed. by John H. S. Burleigh (Philadelphia: Westminster Press, 1953).
5. St. Thomas Aquinas, *Summa Theologica*, I, qu. 2, art. 3; Reproduced in *St. Thomas Aquinas: Basic Writings*, ed. by A. C. Pegis (New York: Random House, 1945), 1:22–23.

6. Martin Luther, *On the Bondage of the Will*, trans. by J. I. Parker and A. R. Johnston, in *Martin Luther*, ed. by John Dillenberger (Garden City: Doubleday, 1961), 181.
7. Ibid., 191.
8. Soren Kierkegaard, *Concluding Unscientific Postscript*, trans. by D. F. Swenson and W. Lowrie (Princeton: Princeton University Press, 1941), 194–95.
9. Kierkegaard, *The Sickness Unto Death*, trans. by W. Lowrie (Princeton: Princeton University Press, 1941), 107–15; 156–63.
10. Kierkegaard, *Concluding Unscientific Postscript*, 182.
11. George Orwell, *Nineteen Eighty-Four* (New York: New American Library, 1981), 176–77.
12. Baruch de Spinoza, *The Ethics*, Part I, def. VI; prop. XIX; prop. XIII; prop. XI; prop. XVI; prop. XVI, cor. III.
13. Edwin Lewis, *The Creator and the Adversary* (New York: Abingdon-Cokesbury, 1948), 20, 138.
14. Paul Tillich, *Systematic Theology*, 3 vols. (Chicago: University of Chicago Press, 1951), 1:59–66.
15. Ibid., 1:14.
16. Ibid., 1:49.
17. Tillich, *The Courage To Be* (New Haven: Yale University Press, 1952), 40–63.
18. Tillich, *Systematic Theology*, 1:186–210.
19. Tillich, *Love, Power, and Justice* (London: Oxford University Press, 1954), 107; *Systematic Theology*, 1:264–65.
20. Tillich, *Biblical Religion and the Search for Ultimate Reality* (Chicago: University of Chicago Press, 1956), 82–83.
21. Tillich, *Systematic Theology*, 1:239.
22. Ibid., 1:49.
23. Ibid., 2:114.

Conclusion

The analogical extensions of Bohr's principle of complementarity to the disciplines of biology, psychology, and theology are ineffective. It has been argued, moreover, that the principle of complementarity is ineffective within quantum theory itself.

Calculations at the theoretical level suffice for the prediction of experimental results. Pictures that describe what happens between observations are irrelevant to this task. The successful prediction of the results of observations is by far the most important goal of the physicist. Pascual Jordan declared that the physicist should

> check most rigorously whether all our assertions, suppositions and problems fit completely into a system of pure description of observed results. Any statement that falls beyond this limit—especially any attempt to express something about the so-called "essence of physical things"—must be eliminated and declared basically meaningless.[1]

The complementary pictures advanced by Bohr refer to explicit experimental arrangements, and are not put forward as descriptions of "particles in themselves" or "waves in themselves." Nevertheless, on Jordan's positivist position, pictorial-level interpretations of quantum systems between observations are to be excluded from physics.

Why should physicists be required to superimpose mutually exclusive causal accounts and spatiotemporal descriptions upon quantum systems during intervals between observations? Bohr supported the claim that complementary pictorial-level interpretations are necessary for the explanation of quantum phenomena by invoking a correspondence argument. This

> correspondence argument . . . expresses the endeavor of utilizing to the outmost extent the concepts of the classical theories of mechanics and electrodynamics, in spite of the contrast between these theories and the quantum of action.[2]

If this is an appeal to the correspondence principle, it is unconvincing. As a methodological directive, the correspondence principle requires that calculations from a theory in a new domain asymptotically approach calculations from a theory that has achieved success in the original domain. Interpreted loosely in this way, the correspondence principle requires that calculations from a theory of the quantum domain asymptotically approach calculations from classical electrodynamic theory for interactions "on a scale sufficiently large to allow the neglect of the quantum of action."[3] However, an asymptotic agreement of calculations, under conditions for which the quantum of action is negligible, provides no support for the claim that complementary causal and spatiotemporal accounts should be given for quantum systems between observations.

Henry Margenau's "Constructivist" View of Quantum Mechanics

Henry Margenau defended this position. He criticized Bohr for recommending that physicists accept an irreconcilable dilemma, the horns of which are complementary "descriptions" in terms of classical observables and causal accounts in terms of abstract states.[4] Margenau maintained that physicists have chosen, and should choose, the second alternative. Nothing is lost by dispensing with complementary pictures of what happens between observations made of a system.

Margenau did not accept Jordan's positivist position. He attributed physical reality to the Born interpretation that correlates $|\Psi|^2\, dV$ with the probability that a subatomic particle is in volume element dV.[5] Schrödinger also attributed objective reality to the Ψ-function. However, Schrödinger assigned reality to Ψ, as a term of the calculus of quantum theory, namely, a wave in multidimensional Hilbert space. Margenau, on the other hand, assigned reality to probability distributions that characterize ensembles of subatomic particles.

Margenau maintained that "physical reality is the class of all verifacts."[6] Verifacts are constructs that (1) participate in circuits of empirical confirmation and (2) satisfy certain methodological requirements. According to Margenau, a circuit of confirmation begins in the data of immediate experience, crosses over into a field of constructs, and returns again to the data of immediate experience.

In his various publications, Margenau created a number of diagrams to represent the confirmation process. A basic feature of these diagrams is the distinction between the P-plane and the C-field.

The P-plane comprises facts that are independent of people's control. Margenau declared that

> a fact is spontaneous in our experience. . . . It is often in the form of an immediate perception or sensory datum, or an observation. . . . As examples I offer the seeing of this shape called desk, the hearing of a sound, the awareness of a pain, or indeed that combination of many such immediacies which we call an *observation* in science.[7]

Among the observations important in science are primary experimental data, such as pointer positions on scales, counter clicks, color changes, the location of the meniscus of a liquid in a glass tube, and the visual impression of a curved path in a Wilson cloud chamber. These data of immediate experience are important to science because, unlike dreams, hallucinations, and illusions, they can be linked to scientific constructs.

Figure 19
A Circuit of Empirical Confirmation[8]

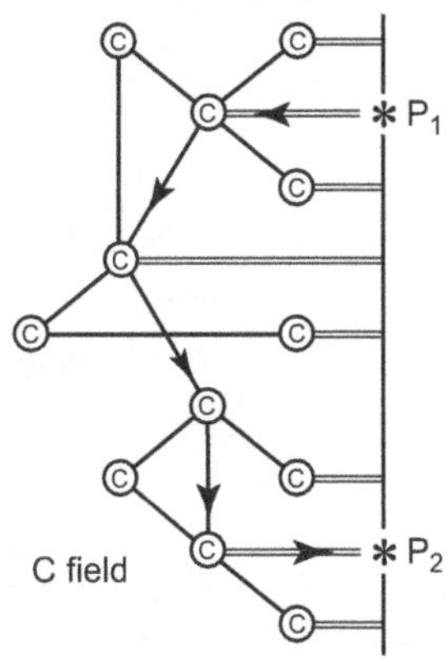

Margenau provided an example of the confirmation process. P_1 represents observations of the position of a comet in the sky. The double-lined arrow from P_1 is a rule of correspondence that enables a translation to be made from observed positions in the sky to a coordinate system in which values of position, velocity, and acceleration are assigned. Rules of correspondence establish connections between sensed nature and constructs.[9]

The single-lined arrows in the C-field represent connections among the concepts of Newtonian mechanics. The physicist utilizes these conceptual connections to establish a transition from the C-field to the P-plane. P_2 represents a prediction of the position in the sky at which the comet will be observed at some other time.

The circuit of confirmation is also a template for scientific explanation. An effective explanation for the observation that P_2 is the case is to trace the path from P_1 through the C-field back to P_2. Margenau declared that

> every physical explanation involves a circuit which starts and ends in two different elements of nature.[10]

Margenau maintained that reification and operational definitions are important rules of correspondence that enable translations to be made from the P-plane to the C-field. These types of rules of correspondence are illustrated in Margenau's diagram of the relation between quantum theory and experience.

Figure 20
The Ψ-Function and the P-plane[11]

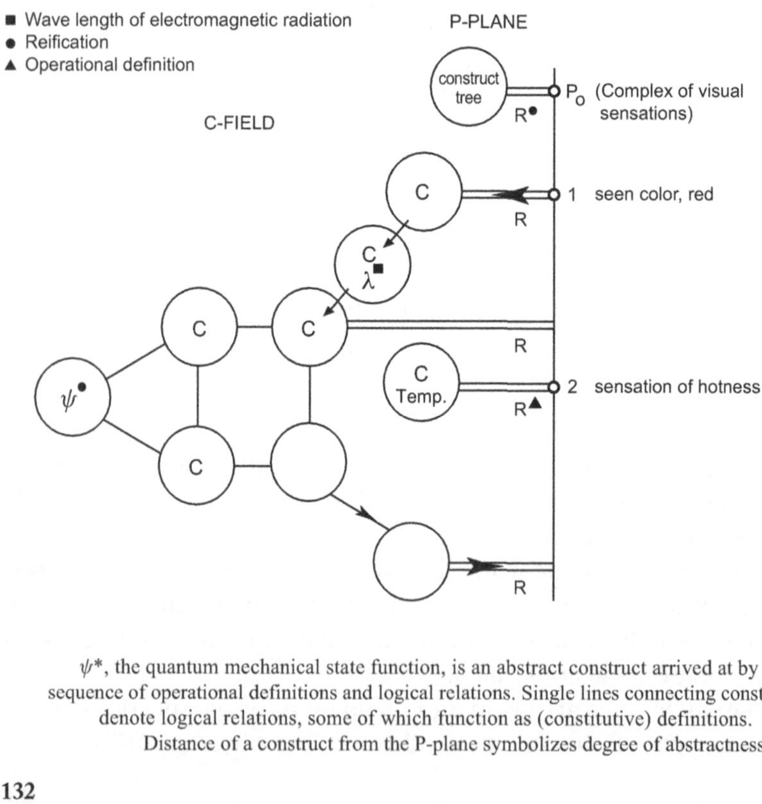

ψ*, the quantum mechanical state function, is an abstract construct arrived at by a sequence of operational definitions and logical relations. Single lines connecting constructs denote logical relations, some of which function as (constitutive) definitions. Distance of a construct from the P-plane symbolizes degree of abstractness.

At the top of the diagram a double-lined arrow indicates a reification of various visual and tactile sensations to form the construct tree. Lower down in the diagram, the concept of temperature is correlated with primary experimental data by an operational definition.

Margenau agreed with P. W. Bridgman that operational definitions provide important links between constructs and the results of instrumental procedures.[12] The form of an operational definition is

$$(x)\ [Ox \supset (Cx \equiv Rx)],$$

where *Oa*—operation *O* is performed in instance *a*,
Ca—concept *c* applies in instance *a*, and
Ra—result of operation *R* occurred in instance *a*.

Various instrumental procedures may be utilized to introduce the concept of temperature into the C-field. In one such procedure, a thermometer is inserted in a fluid. If this operation is performed, then the temperature of the fluid is 70 degrees if, and only if, the thermometer's meniscus is on the "70" mark.

Two additional constructs are identified in figure 20: wavelength λ and the Ψ-function. Wavelength λ is linked via operational definitions to measuring procedures, Ψ is not. Margenau held that distance from the P-plane is a measure of the abstractness of constructs. On this measure, Ψ is a more abstract construct than λ.

Participation in multiple circuits of empirical confirmation is one of two requirements for a construct to qualify as a verifact. The Ψ-function is an abstract concept, but it is linked to concepts, such as charge densities and scattering distributions, and ultimately to data from the P-plane. Ψ satisfies the first requirement above.

The second requirement for status as verifact is that the construct satisfy several regulative principles. In an essay published in 1941, Margenau listed formal implication, permanence, extensibility, simplicity, and causality.[13] In *The Nature of Physical Reality* (1950), he discussed these requirements in more detail.

The requirement of formal implication restricts "concepts admissible in science" to those that are multiply connected to other concepts. "They may not be insular or peninsular; sets forming an island universe must be excluded."[14]

The requirement of permanence restricts verifacts to concepts that retain "integrity of meaning" in all applications.[15] The requirement of extensibility restricts verifacts to concepts open to incorporation in additional theories. Verifacts are open both to the possibility of additional operational links to the P-plane and to the possibility of further links to other concepts.[16]

The requirement of simplicity is notoriously vague. A verifact should play a role in one or more theories that are not unduly complex. Margenau conceded that, although examples can be cited for which one theory is simpler than another, the requirement of simplicity is "more elusive and more difficult to state than all the others."[17]

The requirement of causality stipulates that physical systems be described by theories that imply laws of causal form. Margenau accepted the regularity-between-states theory of causality. A law of causal form specifies the time rate of change of states of a physical system. The mathematical form of such a law is a differential equation that does not contain the time variable in explicit form.

The Ψ-function, on the Born interpretation, participates in numerous circuits of empirical confirmation and satisfies all the methodological principles listed by Margenau. In particular, the theoretical level of interpretation in quantum theory applies to physical systems that are "self-unfolding in a determinate manner," thereby satisfying the requirement of causality.[18] The Ψ-function thus qualifies as a verifact, which means, according to Margenau, that it is an element of "physical reality."[19]

However, there remain difficulties for the "causal attribution implies explanation" thesis, even if one elects to remain at the theoretical level of interpretation. Whereas the behavior of an ensemble of particles may be explained satisfactorily, the behavior of an individual particle is not subject to explanation.

Despite Margenau's protest, many scientists in the 1930s were receptive to the idea that pictures of classical particles and waves are needed to complete the interpretation of quantum phenomena. Quantum mechanics was the first successful theory to deny that phenomena can be explained by unified accounts that specify both causal interaction and spatiotemporal description. There was considerable support for Bohr's attempt to anchor the new theory in the concepts of classical physics.

The situation for a twenty-first century scientist is quite different. A series of highly abstract theories have been advanced recently to account for phenomena in the high-energy–short-time interval domain. For example M-theory features eleven space-time dimensions, and one version of string theory features twenty-six space-time dimensions. It no longer is plausible to require of a high-level theory in physics that it include superimposed pictures of classical waves and particles.

Notes

1. Pascual Jordan, *Physics of the Twentieth Century* (New York: Philosophical Library, 1944), 46.
2. Niels Bohr, "Address at Twentieth-Fifth Class Reunion" (1928). Cited by Niels Blaedel in *The Life of Niels Bohr* (Madison, WI: Science Tech Publishers, 1988), 179.
3. Bohr, "Unity of Knowledge" (1954), in *Atomic Physics and Human Knowledge* (New York: John Wiley & Sons, 1958), 74.
4. Henry Margenau, *The Nature of Physical Reality* (New York: McGraw-Hill, 1950), 421–22.
5. Margenau, "Metaphysical Elements in Physics," *Rev. Mod. Phys. 13* (1941), in *Physics and Philosophy: Selected Essays* (Dordrecht: Reidel, 1978), 108; "Philosophical Problems Concerning the Meaning of Measurement in Physics," *Phil. Sci. 25* (1958), in *Physics and Philosophy: Selected Essays,* 209.
6. Margenau, "Reality in Quantum Mechanics," *Phil. Sci. 16* (1949), 287–302.
7. Margenau, *Open Vistas* (New Haven: Yale University Press, 1961), 5.
8. Ibid., 16.
9. Margenau, "Metaphysical Elements in Physics," in *Physics and Philosophy: Selected Essays,* 110.
10. Ibid., 99.
11. Margenau, "Religious Doctrine and Natural Science," in *Physics and Philosophy: Selected Essays,* 375.
12. Margenau, *The Nature of Physical Reality,* 231–32.
13. Margenau, "Metaphysical Elements in Physics," 110.
14. Margenau, *The Nature of Physical Reality,* 87.
15. Ibid., 90.
16. Ibid., 90–94.
17. Ibid., 96–98.
18. Margenau, "Advantages and Disadvantages of Various Interpretations of the Quantum Theory," *Physics Today 10* (1954), 13.
19. Margenau, "Reality in Quantum Mechanics," 287–302.

Index of Names

Index of Subjects

For Product Safety Concerns and Information please contact our
EU representative GPSR@taylorandfrancis.com Taylor & Francis
Verlag GmbH, Kaufingerstraße 24, 80331 München, Germany